Environmental management in the Dutch food and beverage industry

Environmental management in the Dutch food and beverage industry

A longitudinal study into the joint impact of business network and firm characteristics on the adoption of environmental management capabilities

Derk-Jan Haverkamp

Innovation and sustainability series – Volume 1

Wageningen Academic
P u b l i s h e r s

ISBN: 978-90-8686-050-0
ISSN: 1874-7663

First published, 2007

Wageningen Academic Publishers
The Netherlands, 2007

The individual contributions in this publication
and any liabilities arising from them remain the
responsibility of the authors.

The publisher is not responsible for possible
damages, which could be a result of content
derived from this publication.

Table of contents

Acknowledgements

The research reported in this book was carried out at the Management Studies Group of Wageningen University and Research Centre (WUR). My sincere acknowledgements go to all who contributed to the completion of the work.

Parts of the project were made available through financial and supervisory support from both the Dutch Ministry of Housing, Spatial Planning, and the Environment (VROM) and the Dutch Agri Chain Competence centre (ACC), which are gratefully acknowledged. Furthermore, I would like to thank all respondents from the Dutch food and beverage firms that participated in the research. In addition, the willingness of the interviewees to share their environmental experiences with me is highly appreciated.

My special thanks go to Onno Omta for a warm welcome at the Management Studies Group. Onno, I appreciate your pragmatic guidance and persistence to improve the quality of the academic reporting. I also want to express my special thanks to Harry Bremmers for the daily supervision and support, including hours of sharing all kinds of problems. Harry, I truly admire your multi-disciplinary expertise and ease of reasoning.

I also acknowledge the encouragement received from Prof. Metz to extend my academic background with Management Studies. It turned out to be a significant shift in my academic background, which I truly appreciate. Furthermore, I would like to mention that the philosophical lectures on 'nature versus environment' by Prof. Wissenburg have importantly contributed to my academic inspiration.

Willemijn van Dusseldorp and Arjan Piëst are acknowledged for carrying out interviews and their work on the quantitative data. Furthermore, I am indebted to Mijke Smit for gathering parts of the data presented in this book.

The good contacts with the staff of the Management Studies Group were essential to me. I would like to thank them for showing a critical attitude as well as their friendship during coffee breaks, drinks and department trips. In particular, I would like to thank Ron Kemp for his support with the statistical analysis. In addition, I would like to express my sincere acknowledgement to Geoffrey Hagelaar and collegue PhD-students for collectively organizing the capita selecta course *Innovation in Business Administration*.

I would like to thank my colleagues at ENVIRON for a pleasant working atmosphere and the time provided to finalize this book.

I. Introduction

In recent years, people worldwide have become increasingly aware that environmental pollution is seriously affecting the living conditions of contemporary society and, if nothing is done, future generations. Negative effects on the natural environment (hereafter called 'the environment') have been extensively reported in the media. Illustrative is the documentary *An inconvenient truth* in which the negative environmental effects of global warming are exposed. Global warming has severe consequences for various ecosystems and threats of increasing temperature include a rising sea level that has important implications for countries that are geographically situated below sea level, like the Netherlands (Root *et al.*, 2003; PCCC, 2006). Interestingly, the importance of taking care of the environment was stressed during the last G8 summit of the world's eight richest industrial countries (Germany, UK, France, Italy, Canada, USA, Japan, and Russia). It was agreed to work on a post-Kyoto Protocol to reduce the emission of greenhouse gases.[1] At the European and national levels too, increased environmental attention can be seen. The *Future Environmental Agenda* of the Dutch government consists of different mandatory EU emission reduction goals to improve air, water, and soil quality. In addition, the Dutch government has formulated environmental targets, such as those to reduce noise hindrance and waste (VROM, 2006a).

This book focuses on the attention firms give to reducing their environmental impact. Industry is responsible for environmental emissions that contribute negatively to air, water, and soil quality. As part of a longitudinal study, the Arthur D. Little Innovation High Ground Survey of 40 managers of technology firms across Europe, U.S.A. and Japan showed that environmental awareness has grown considerably over the past few years (Hedstrom *et al.*, 2005). The managers indicated that sustainability-driven innovation has a growing potential to deliver value to business. However, only 5% indicated paying attention to environmental issues in their strategic planning and decision-making. As the survey was conducted among large multinationals (e.g. Sony, Procter & Gamble, Vodafone, HP, Motorola, and Dupont), these figures would probably look even worse if small- and medium-sized firms were included (see also: Hillary, 1999; Del Brío and Junquera, 2003). This conclusion is underpinned by critiques of a rather defensive environmental strategy adhered to especially by small and medium-sized Dutch firms, but also by many large Dutch firms, despite growing attention for the environment in national and international markets.[2] *This book aims for a deeper understanding of the factors that have an impact on the adoption of environmental management capabilities in firms in the Dutch food and beverage industry.*

The Dutch food and beverage industry is responsible for significant pressure on the environment (Dutilh and Blijswijk, 2004). An important environmental issue in this industry is water usage for washing fresh products (potatoes, sugar beets, etc.) and cleaning of the production

---

[1] *Volkskrant*, June 9 2007.

[2] *NRC*, June 1 2007.

process and equipment. It accounted for about 70% of the phosphate emissions to water by the Dutch industry in 2004. Substantive amounts of energy are consumed for heating (e.g. frying and baking) as well as cooling (freezing and storage of products). Because many food and beverage products are suited to the consumer market, which implies that their batch size is small, it is also responsible for about 60% of all packaging material. Fortunately, much has been done to reduce it, while at the same time about 60% consists of recycled materials (e.g. glass and paper). Furthermore, the food and beverage industry produces organic waste. For example, slaughterhouses have to deal with the material that remains after the animals have been slaughtered (bones, skin, blood, etc.). Vegetables and fruit processors select products which are suitable for the consumer market, while rotten products are sold to animal feed producers. Lately, there has been a growing interest in the use of organic waste to generate bio-fuels, which can be used at the manufacturing site or sold on the market (SenterNovem, 2005). In total, 90% of organic waste was recycled in 2003. The substantive pressure on the environment, on the one hand, and the attention to reducing the environmental impact on the other, makes it interesting to take a closer look at the factors that influence the attention for environmental management in the Dutch food and beverage industry.

Academic interest in environmental management has emerged from strategic management (e.g. Roome, 1994; Clarkson, 1995; Porter and Van der Linde, 1995; Hart, 1997; Porter and Kramer, 2006), management and organization (e.g. Cramer, 1998; King, 2000; Prakash, 2000; Kolk and Mauser, 2002), marketing (e.g. Polonsky, 1995; Prakash, 2002; Peattie and Crane, 2005), financial management and accounting (e.g. Schaltegger *et al.*, 1996; Kolk, 2000), as well as operations and logistics (e.g. Gupta, 1994; Bloemhof-Ruwaard *et al.*, 1995). These disciplines have each highlighted different aspects, such as the strategic importance of taking care of the environment, the operational consequences of clean technologies, the financial benefits of reducing emissions, and environmental marketing opportunities. Cramer (1998) states that *environmental management in theory involves the study of all technical and organizational activities aimed at reducing the environmental impact caused by a company's business operations*. However, the author herself criticizes this definition for being inconclusive, since it also includes unintended pollution reduction, while the firm's strategy may lack attention for environmental issues. This book focuses on the commitment to the environment in firms in terms of the adoption of environmental management capabilities, which comprise different environmental management items, such as an environmental action program, regular environmental auditing to evaluate strategic environmental targets, and an environmental database to keep a record of environmental performance. In short, they reflect the capacity of the firm to take care of the environment on a structural basis.

A recent contribution by Porter and Kramer (2006) on the interdependency between industry and society serves as a point of departure for the conceptual framework, which integrates the outside-in and inside-out perspective on the adoption of environmental management capabilities. The outside-in perspective originates from the industrial organization theory, which evaluates the influence of industrial forces, such as rivalry among competitors and entrance of new market parties, on the firm's competitiveness (Porter, 1980; Porter, 1981). In line with this view, the network theory stresses the interrelatedness between a firm and other actors, like buyers, suppliers, and competitors, in its business network (Håkansson and Snehota, 1989; Powell, 1990; Omta *et al.*, 2001). They can impose requirements on the firm to take care of the environment. But other environmental stakeholders, like local inhabitants and environmental groups, can also execute pressure to reduce noise and smell hindrance on a local level. The inside-out perspective is embedded in the competence view. It evaluates the firm's competitiveness from acquiring valuable resources, competences, and (dynamic) capabilities to deal with these external influences (Barney, 1986b; a; Prahalad and Hamel, 1990; Teece *et al.*, 1997). It examines the importance of tangible assets, like production technologies and equipment, as well as intangible assets, such as knowledge and skills, for achieving strategic business goals including a reduction of the environmental impact.

Multiple authors emphasize that firms can create competitive benefits from taking care of the environment by the reduction of environmental pressures, such as those related to less waste disposal and more efficient production efficiencies, but also environmentally friendly product differentiation, including environmental orientation of product features (e.g. recyclable components) and packaging (Hart, 1995; Porter and Van der Linde, 1995; Shrivastava, 1995; Adner and Helfat, 2003). From this perspective, Hart (1995) discusses the natural resource-based-view (natural-RBV) in his well-known seminal paper. He evaluates the competitive advantages of pollution prevention, product stewardship, and sustainable development capabilities. Pollution prevention capabilities aim to reduce environmental emissions, using clean technologies: they can enhance an environmentally responsible reputation, which is essential to guarantee a social license-to-produce (Kagan *et al.*, 2003). Product stewardship refers to environmental attention from the perspective of the total product-life-cycle from *cradle-to-grave*, or *cradle-to-cradle*, if waste outflows can be re-used (McDonough and Braungart, 2002). Competitive benefits can be gained from environmentally responsible product (re)design practices to meet the *green* expectations of customers. Last, sustainable development capabilities refer to collective efforts to make industry less dependent on depleting natural resources (Hart, 1995; 1997). From this perspective, environmental cooperation with other chain actors can be interpreted as an important step towards developing sustainable management capabilities.

Acceptance by stakeholders and thus business continuation increasingly depends on the firm's capability to satisfy, not only economic, but also social and ecological stakeholder wishes (Elkington, 1998). The challenge for firms is to stay socially legitimate, while remaining profitable. The adoption of environmental management capabilities will be essential in this context. However, their development is not a naturally enrolling process, since it requires learning to adapt the organization to new environmentally friendly working routines at both the strategic and operational level (e.g. Cramer, 2005). In addition, coming up to different stakeholder expectations will not be easy, because they are often diverse and potentially competing (Mitchell *et al.*, 1997). Stakeholder wishes will not necessarily be in line with business priorities. This study tries to determine the trade-off between business network and firm characteristics on the adoption of environmental management capabilities.

Figure 1.1 shows the research framework. A distinction is made between stakeholder pressures and environmental cooperation in the business network. The stakeholders can be divided into three groups: government, chain and network actors (e.g. buyers and suppliers, branch-organizations, and financial institutions), and societal groups (e.g. environmental organizations and the local community). Previous studies have shown that stakeholders can exert pressure on firms to pay attention to the environment (e.g. Henriques and Sadorsky, 1999; Sharma and Henriques, 2005). Although government often has the most influence, because it can exercise coercive pressure to comply with environmental regulation, firms perceive a strong impact from other stakeholders as well (Braglia and Petroni, 2000; Buysse and Verbeke, 2003; Sharma and Henriques, 2005). Clarkson (1995) discerns primary and secondary stakeholders. Primary stakeholders (e.g. buyers and suppliers) are of vital importance for the achievement of the primary business goals, whereas secondary stakeholders (e.g. societal groups) can only influence primary goal attainment. Government occupies a special position, because it can adopt the characteristics of both primary and secondary stakeholders. It has environmental interests in the firm from a legal point-of-view and it can increase the attention for environmental issues by providing negative and positive incentives. Negative incentives refer to financial sanctions and the withdrawal of the legal *license-to-produce* (i.e. governmental environmental permits), if a firm acts in non-compliance with environmental regulation (Kagan *et al.*, 2003). By contrast, positive incentives include support, like subsidies for investments in clean production technologies or environmental cooperation in terms of public-private voluntary agreements between government and firms (or: covenants). Government can, for instance, provide environmental feedback to firms that are participating in covenants to improve environmental performance.

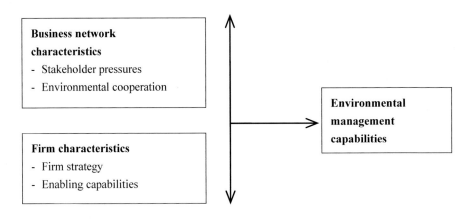

Figure 1.1. Research framework linking outside-in and inside-out approaches to the adoption of environmental management capabilities.

Firms do not operate in isolation, but work together in chains and networks (Omta *et al.*, 2001). The chain actors include buyers, suppliers, and consumers. While the environmental wishes of consumers will be an important motive to pay attention to the environment (Prakash, 2002; Peattie and Crane, 2005), firms are also increasingly confronted with environmental requirements in business-to-business relationships, because buyers have to fulfil the environmental wishes of their customers. This can be referred to as the *green multiplier effect* (Holt, 2004). Furthermore, buyers and suppliers have an interest in adequate control of environmental emissions at the chain level in order to reduce the chance of interruptions or delays in supplied materials due to firms with environmental problems, such as illegal pollution activities which can result in (temporary) closure of facilities (Lippmann, 1999). However, they might not merely impose environmental requirements, but also look for cooperative efforts to reduce environmental impacts at the chain level, for instance with respect to the use of environmentally friendly raw materials or recyclable packaging. The network actors include facilitating organizations, such as branch-organizations which can provide support for the implementation of environmental activities in terms of environmental information and a discussion platform for firms to exchange environmental experiences. Banks and insurance firms are also part of the network. They may require the implementation of environmental activities conditional to the provision of their services, such as loans and insurance.

The firm characteristics include firm strategy and enabling capabilities (see Figure 1.1), such as experience with the implementation of food quality and safety systems and environmental knowledge of employees that can be used for the adoption of environmental management capabilities. Notably, this book uses the concepts of capabilities and competences as interchangeable, since they are closely related to each other (see Section 4.3.1). Aragón-Correa (1998) found that a prospector strategy, which focuses on the implementation of innovative technologies, could significantly explain, in a strictly statistical sense, the attention for environmental management capabilities. This can be understood from the fact that the business strategy forms an important basis for the environmental strategy. Firms with a cost-minimization strategy will, for instance, be interested in reducing pollution as long as it contributes to cost savings, while firms with a product differentiation strategy might be willing to invest in more advanced clean production technologies. Interestingly, Christmann (2000) found that the cost advantages of taking care of the environment were moderated by capabilities related to a prospector strategy, such as the search for and experience with the implementation of new production technologies. The advantages were cost savings that complied with environmental regulations and (consequently) an improved competitive position. Other studies also established a positive trade-off between various existing organizational capabilities and attention for environmental management capabilities (Melnyk *et al.*, 2003; Sroufe, 2003; González-Benito and González-Benito, 2005).

Dynamic capabilities can be interpreted as a subset of organizational capabilities that are needed to respond to changing market demands (Teece *et al.*, 1997; Eisenhardt and Martin, 2000). In this context, the present study evaluates the attention for environmental capabilities from a dynamic capabilities perspective, since it examines the stakeholder influences that are essential to take into account for business continuation (e.g. pressure from customers to pay attention to the environment and environmental requirements induced by government). In addition, a longitudinal analysis is provided based on data gathered in 2002 and 2005. A longitudinal assessment that was carried out gives insight into changing stakeholder pressures that are important to anticipate. Referring to the research framework (see Figure 1.1), the central research question that will be answered in this book is formulated as follows. *What is the joint impact of the business network (government, buyers, suppliers, customers, etc.) and firm characteristics on the adoption of environmental management capabilities in Dutch food and beverage firms?*

Environmental management capabilities are evaluated by investigating the implementation of different elements of an environmental management system. The latter can be defined as part of the overall management system that aims to reduce the environmental impact based on the principle of continual improvement of quality in the Deming plan-do-check-act cycle (Netherwood, 2004). International standards and guidelines have emerged on the implementation of environmental management systems, such as ISO14001 and EMAS, which focus on environmental activities: the formulation of an environmental action program, training and education of personnel, regular measurement and registrations of emissions, internal environmental information processing, etc. (Starkey, 2004). Furthermore, the present study evaluates environmental management capabilities that can be used to reduce environmental impact along the product-life-cycle, such as environmental information exchanges with chain partners (e.g. buyers and suppliers) and environmentally responsible product (re)design to lower environmental emissions associated not only with production processes, but also product characteristics.

The remainder of this book is structured as follows. Chapter 2 gives an overview of previous studies. It elaborates on the academic interests in environmental management and different ways in which it is operationalized. It also evaluates previous empirical studies that approached environmental management from the outside-in perspective by considering influences from different stakeholder from the business network. Furthermore, attention is paid to studies that adopted an inside-out perspective on the development of environmental management capabilities by looking at the importance of firm strategy and organizational capabilities in terms of improved business performance.

Chapter 3 focuses on (inter)national environmental policy developments. The Dutch governmental environmental policy is reviewed in particular. In time, it has been shown to move away from centrally formulated policy making towards more participatory arrangements with industry. In addition, this chapter discusses the economic and environmental characteristics of the Dutch food and beverage industry. It shows that it represents a vital part of the Dutch economy, while it is also responsible for significant emissions to air, water, and soil. The attention to environmental and food quality issues in the different food and beverage sectors is also discussed. This chapter evaluates the implementation of different management standards and guidelines (e.g. HACCP and ISO9001), as well as the involvement of branch-organizations and product boards in environmental issues.

Chapter 4 presents the theoretical background. It reviews the outside-in perspective by discussing the concept of network embeddedness paying special attention to the position of the firm in the supply chain (e.g. Granovetter, 1985; Coleman, 1990; Burt, 1992; Uzzi, 1997). Furthermore, it evaluates different stakeholder typologies (Clarkson, 1995; Mitchell *et al.*, 1997). The inside-out perspective is reviewed by focusing on organizational resources, competences, and dynamic capabilities (e.g. Prahalad and Hamel, 1990; Teece *et al.*, 1997; Barney, 2001). Different firm strategies to lower environmental pressure, using the natural resource-based view (RBV), are discussed (e.g. Hart, 1995; Litz, 1996).

Chapter 5 concentrates on the study design, which comprises the conceptual model, the propositions, and the operationalization of the research variables. The present study employs a mixed study design, since it combines semi-structured interviews involving 13 environmental managers of different food and beverage firms together with two large-scale surveys. The surveys are used to quantitatively measure the impacts of the business network on the adoption of environmental management capabilities in 2002 and 2005. Micro (5-10 employees), small (10-50), medium (50-250), and large (>250 employees) firms are included in 2002. In total, 492 questionnaires could be used for the analyses, of which 386 originated from micro and small and 106 from medium-sized and large Dutch food and beverage firms. In 2005, special attention is paid to firm strategy and enabling environmental capabilities (see Figure 1.1). The survey included medium-sized and large firms only to get a more homogenous sample. In total, 100 questionnaires could be used. The data analyses are carried out using univariate and multivariate statistical techniques on the separate databases. It also comprises a longitudinal analysis to get insight into changes with respect to business network impacts. This is relevant, as it is expected that the impact of different stakeholders has grown in recent years under the influence of the increased societal attention for the environment. Kagan *et al.* (2003) found, for instance, that not only government influence, but also growing pressure from local communities and environmental organizations are important motives for firms to clean up.

Chapter 6 discusses the results of the semi-structured interviews. It evaluates the main environmental characteristics of the interviewed firms. Furthermore, the different elements of the research framework are discussed, such as the importance of the relationship with government in terms of perceived environmental support and the quality of environmental information provided by government. Attention is paid to other stakeholders too, such as chain and network actors and societal groups. In addition, the interviews deal with strategic and operational environmental issues, including top-management commitment and the development of environmental management capabilities.

Chapter 7 presents the results of the surveys. It discusses the sample characteristics of both surveys, including the response rates, background of the respondents and a non-response analysis. It assesses the results of a descriptive analysis on the attention paid to environmental management capabilities in the Dutch food and beverage industry, as well as the perceived business network impacts. The importance of firm strategy and enabling environmental capabilities are evaluated. The correlations between the research variables and the results of the regression analyses are evaluated. Organizational configurations are compiled based on a cluster analysis. The empirical results of the longitudinal analysis are presented. They discuss the changes in the perceived stakeholder influences on environmental management in the medium-sized and large food and beverage firms in 2005 compared to 2002.

Chapter 8 discusses the theoretical and managerial implications. It evaluates the propositions and it draws conclusions with respect to the academic contribution of the present study, as well as the possibilities for further research. This chapter ends with a discussion on the managerial implications for Dutch food and beverage firms, as well as governmental environmental policy.

2. Overview of previous studies

This chapter focuses on previous empirical research that is relevant to the present study. Section 2.1 starts with a discussion on approaches towards environmental management found in literature. These approaches concentrate on sustainable production, corporate social responsibility (CSR), ISO14001 certification and 'best practices' of environmental management. Section 2.2 reviews the quantitative studies that took an outside-in perspective on the adoption of environmental management capabilities by considering stakeholder influences from the business network. Section 2.3 proceeds with a review of the empirical studies that adopted an inside-out perspective on attention for environmental management capabilities by evaluating the managerial and financial consequences of paying attention to the environment in relation to firm strategy and/or organizational capabilities. Section 2.4 provides the concluding remarks with respect to this chapter, including a discussion on the novelty of the present study.

2.1 Approaches towards environmental management

The World Commission on Environment and Development (1987) had put 'sustainable production' on the societal agenda by the end of the eighties and formulated it as *the development that meets the needs of the present without comprising the ability of future generations to meet their own needs.* Although it is still widely applied, other different definitions have emerged since then. They stress one or more of the following issues: inclusiveness, connectivity, equity, prudence, and security (Gladwin *et al.*, 1995). Inclusiveness refers to the holistic character of the concept: sustainable production encompasses not only consideration of human, but also environmental values, at the local and global level, at present and in the future. Connectivity refers to the interdependency of different environmental and social problems, such as between scarcity of environmental resources and poverty. Because of this, Welford (1998; 2000) stresses that firms have a moral obligation to consider human and ecological systems. Equity refers to a desired equal distribution of resources and property rights both in present and future generations and prudence reflects the avoidance of irreversible damage to ecosystems. Last, security implies safeguarding the living conditions of contemporary and future generations. In conclusion, sustainable production includes multiple dimensions, which makes it difficult to implement (e.g. Holmberg and Robèrt, 2000; Robèrt *et al.*, 2002).

An important step towards sustainable production is the growing attention for corporate social responsibility (CSR). The European Commission (2001) states that CSR is *a concept whereby companies integrate social and environmental concerns in their business operations and in their interaction with their stakeholders on a voluntary basis.* It includes a normative component, since it refers to voluntary initiatives by firms to take care of their impact on society and the environment. Elkington (1998) makes a linkage to economic motives stressing that firms are challenged to consider social and environmental issues, while generating profits. This has become known as the triple-bottom-line approach, consisting of the people, planet, and profit dimension (Elkington, 1998; Epstein and Roy, 2001). The people dimension refers to consideration of stakeholder interests in the organization, such as employees that are concerned about safe working conditions. Stakeholders from outside the firm are, among others, societal groups that wish for a reduction in local environmental emissions and other chain actors (e.g. buyers and suppliers) that are interested in safeguarding the supply of materials. The planet dimension implies taking care of the environment. The profit dimension stresses the need to establish a profitable balance between taking care of social and environmental issues to safeguard business continuation. On the whole, CSR implies managerial priority for multiple issues, which is not easy to achieve, because of potentially conflicting interests, such as investments in clean technologies that cause a burden on short-term profitability. The consideration of social and environmental issues requires establishing new working routines (Cramer *et al.*, 2004; Cramer, 2005). Hahn and Scheermess (2006) found, based on a cluster analysis, that only 25% of the 195 German firms they studied succeeded in integrating social and environmental issues in their organization, while 42% focused on environmental issues only and 33% did not pay any or limited attention to social and environmental issues. The present study evaluates different organizational profiles as well (see Section 7.4.5).

Several studies have evaluated different environmental management practices. In general, a distinction can be made between defensive (e.g. the implementation of end-of-pipe technologies) and proactive (e.g. using pollution prevention technologies) environmental activities (e.g. Christmann, 2000; Buysse and Verbeke, 2003). Examples of these activities will be discussed in Section 2.2 and 2.3. Furthermore, numerous studies have focused on the implementation of environmental management systems (Quazi *et al.*, 2001; Bansal and Bogner, 2002; Fryxell and Szetto, 2002; Halkos and Evangelinos, 2002; Melnyk *et al.*, 2003; Fryxell *et al.*, 2004; Zutshi and Sohal, 2004a; b). The majority operationalized it using ISO14001 certification, because it reflects the availability of a workable environmental management system (e.g. Fryxell and Szetto, 2002; Halkos and Evangelinos, 2002; Melnyk *et al.*, 2003). One of the exceptions, though, is Sroufe (2003), who defined environmental management system implementation as the sum of different environmental activities, such as information gathering on environmental emissions, the availability of an environmental database, and environmental information processing. A comparable strategy is applied by the present study, which measured the different elements of an environmental management system rather than ISO14001 certification only (see Section 5.3).

2.2 Business network impact

Table 2.1 includes an overview of representative quantitative studies that measured business network impacts on the attention for environmental management capabilities. It shows the literature reference, the measured stakeholder influence, the main results, additional comments on the results, as well as the data source used. The majority of the studies finds that government has the strongest impact on environmental management (Garrod and Chadwick, 1996; Henriques and Sadorsky, 1996; Madsen *et al.*, 1997; Braglia and Petroni, 2000; Madsen and Ulhøi, 2001).

Henriques and Sadorsky (1996) were among the first to show empirically that attention for the environment can be explained by stakeholder pressure (see Table 2.1). They use a logistic regression model with the implementation of a formal environmental plan as the dependent variable. Customers, shareholders, government, and the local community turn out to be positively associated with the presence of an environmental plan. They also asked respondents to indicate the influence of other stakeholders, such as competitors, branch-organizations and insurance firms, which are not pre-defined in the questionnaire. It appears that this group of stakeholders as a whole has, in a strictly statistical sense, a significant negative effect on the implementation of an environmental plan. The authors explain this by pointing to the diversity of interests of the included stakeholders. It would, however, be more interesting to know which stakeholders have a negative impact: this remains uninvestigated.

In another study by Henriques and Sadorsky (1999) a cluster analysis is carried out including six items that can be associated with an environmental management system, such as a written environmental plan, environmental information processing, and top-management commitment. The cluster analysis results in reactive, defensive, accommodative, and proactive environmental management practices (see also Section 4.2.3). The stakeholder pressures are divided into regulatory, community, organizational, and media influences, using factorial data reduction. Differences in mean values on stakeholder pressures are tested with an ANOVA t-test. Next, the stakeholder impacts are included as the dependent variable in a regression analysis: it appears that environmentally proactive firms can be associated with significantly stronger perceived influences from regulatory and organizational stakeholders, but not from the local community and the media.

Madsen and Ulhøi (2001) use factorial data reduction to divide the business network impacts of Danish industrial firms into three categories. The first factor comprises stakeholders with a limited impact on the adoption of environmental management capabilities (e.g. environmental organizations), whereas the other two stakeholder groups have either a direct (e.g. government and employees) or indirect (e.g. competitors and customers) impact on the firm. However, in prior research in the Danish industry, Madsen et al. (1997) found that different stakeholders load on the same factors (i.e. limited, direct, and indirect impacts). Although this might be related to the fact that more stakeholders are included in the analysis of 2001 compared to 1997, which could affect the results of the factor analysis, it may also be related to changes in exerted influences between stakeholders. In conclusion, it emphasizes the need for a longitudinal analysis to get more insight into whether and how business network impacts are changing in time.

Braglia and Petroni (2000) make a distinction between more and less environmentally committed firms based on the availability of elements of environmental management capabilities, such as using recycled materials and environmentally responsible product (re)design. They test significant differences in stakeholder pressures between these two groups. Interestingly, a t-test shows that firms that are more committed to the environment perceive

Table 2.1. Quantitative studies on stakeholder influences on attention for the environment.

Reference	Stakeholder influence	Main results
Garrod and Chadwick (1996)	Perceived stakeholder influence on environmental management	Government is the most influential stakeholder, followed by customers, while the impact of other stakeholders is far less important.
Henriques and Sadorsky (1996)	Perceived stakeholder pressure to consider environmental issues	The formulation of an environmental action plan is positively influenced by customers, shareholders, government, and the local community. A negative influence was measured for lobby groups.
Madsen et al. (1997)	Perceived influence of drivers (stakeholders) behind environmental improvement efforts	Owners, national legislation, and customers are the most important driving forces, followed by trade unions and international regulation. By contrast, financial institutions and consumer organizations are the least important.
Henriques and Sadorsky (1999)	Perceived stakeholder pressure to consider environmental issues	Factor analysis reveals four stakeholder groups: regulatory, community, organizational, and media. Pro-active firms are significantly influenced by all these groups, expect the media. By contrast, reactive firms attribute a strong influence to the media
Braglia and Petroni (2000)	Perceived stakeholder pressure on environmental care	Customer requirements, regulation, and competitors are important drivers of pro-active corporate environmental practices. Committed firms experience strong influence from local community and environmental organizations.
Madsen and Ulhøi (2001)	Perceived influence of different stakeholders on environmental initiatives	Owner/shareholders, local and national authorities have the strongest influence. Also customers and international authorities are influential. Larger firms experience stronger influences.
Buysse and Verbeke (2003)	Perceived stakeholder pressure on decisions related to environmental management	A factor analysis revealed four groups: external primary ($\alpha=0.84$), secondary ($\alpha=0.80$), internal primary ($\alpha=0.72$), and regulatory ($\alpha=0.72$) stakeholders. Firms with a reactive or pollution prevention strategy are influenced by regulatory stakeholders. Pro-active firms by other too.
Zutshi and Sohal (2004b)	The importance of employees and suppliers for environmental management	Employee involvement can contribute to environmental management, while the involvement of suppliers is still very limited
Sharma and Henriques (2005)	The influence of stakeholders on different environmental management practices	Different withholding and usage strategies of different stakeholders are associated with more advanced environmental management practices (like eco-design).

Description/comments	Data source
The analyses are descriptive only.	Survey of 26 U.K. firms with > £2 million turnover (response rate: 15%)
Environmental responsiveness is measured as the implementation of a formal plan for dealing with environmental issues. A logistics regression analysis is performed.	Survey of 400 firms, among the 750 largest in Canada (response rate: 53%)
Factor analysis revealed three groups: stakeholders with limited direct influence on the firm, regulators and internal stakeholders, and market-related stakeholders.	Survey results of 272 Danish firms from multiple industries (response rate: 55%)
See also: Henriques and Sadorsky (1996).	Survey of 400 firms, which are among the 750 largest firms in Canada (response rate: 53%)
Environmental pro-activeness is measured as the attention for the environment at different business activities (R&D, product design, manufacturing, etc.).	Survey of 120 firms from the Italian food packaging industry (response rate: 20%)
Factor analysis revealed three groups: firms with limited, direct, and indirect influence. See also: Madsen et al. (1997).	Survey of 300 Danish industrial firms with at least 10 employees (response rate: 60%)
Firms are clustered in three groups: reactive, pollution prevention, and environmental leadership strategies.	Survey of 197 Belgian industrial firms (response rate: 44%)
Only employees and suppliers were considered as stakeholders.	Survey of 286 firms from Australia and New Zealand (response rate: 50%)
A distinction was made between six environmental management practices ($\alpha \geq 0.63$) based on secondary data, such as environmental reports.	Survey of 49 Canadian forest firms (response rate: 25%)

Table 2.1. Continued.

Reference	Stakeholder influence	Main results
Eesly and Lenox (2006)	Stakeholder salience	Stakeholders with relatively strong power compared to the targeted firm and whose requests are more legitimate have a higher level of influence.

a significantly stronger impact from customers, government, competitors, and societal groups (e.g. the local community and environmental organizations). Unfortunately, the joint impact of different stakeholders in identifying the relative importance of each stakeholder group is not statistically assessed.

Buysse and Verbeke (2003) employ a research methodology similar to Henriques and Sadorsky (1999). A cluster analysis is carried out to compile environmental strategy profiles among Belgium industrial firms. They measured, among other things, the availability of an environmental management system and environmental management procedures. Three profiles emerge from their analysis, namely a group of firms with a reactive strategy, pollution prevention, and environmental leadership strategy. The leadership strategy refers to firms that want to show environmental excellence (see also Section 4.3.3). A comparison of the mean values indicates that firms with environmental leadership can not be associated with significantly stronger impacts from government compared to pollution prevention firms. Accordingly, the authors state that environmental leadership has little to do with managing stakeholder relationships in the regulatory sphere. They therefore suggest that 'command-and-control' based governmental policies may be extended to include cooperative public-private efforts to stimulate firms to go beyond pollution prevention. The effect of this recommendation on attention for the environment is, however, not empirically tested.

Sharma and Henriques (2005) evaluate the influence of stakeholders that try to withhold resources from the firm ('withholding strategies') versus those that aim to make use of the firm ('usage strategies') to satisfy their own interests. Examples of withholding strategies are customers that cancel their orders and/or societal groups that try to disrupt operations through protests at the manufacturing site. Usage strategies refer to buyers that demand product certification to satisfy the wishes of their own customers and/or environmental organizations that provide support to the firm and find remediation against environmental pressures (e.g. environmental knowledge). A factor analysis is carried out. It demonstrates six environmental practices, which can, among others, be differentiated based on attention for energy efficiency, recycling of materials, as well as product life-cycle analysis. The different practices are used as a dependent variable in a regression model. It appears that firms with

Description/comments	Data source
Only actions by secondary stakeholder groups are considered: 333 proxy votes, 144 civil suits, 78 protests, 38 boycotts, and 9 letter-writing campaigns.	Database of 602 environmental actions against U.S. firms during 1971-2003 (LexisNexis Academic database)

more proactive environmental practices are, in a purely statistical sense, more influenced by both the withholding strategies of environmental organizations as well as the usage strategies of customers. Unfortunately, the authors do not evaluate the influence of government. Since prior studies have revealed a dominant influence of government, it would have been interesting to evaluate various governmental withholding and usage strategies, such as pressure based on the withdrawal of environmental permits and public-private environmental partnerships. From this perspective, the present study adds an empirical evaluation of environmental voluntary agreements between government and firms.

2.3 Firm strategy and organizational capabilities

Table 2.2 includes an overview of quantitative studies that adopted an inside-out perspective on the adoption of environmental management capabilities. It shows the references, measured dependent variable(s), the main results, some additional comments, and the data source used. The major concern of most of the studies is to get more insight into the relationship between environmental care and business performance in terms of operations and financial benefits through, for instance, increased production efficiency. In general, empirical evidence confirms a positive relationship, in the purely statistical sense, between the adoption of environmental management capabilities and business performance (Klassen and McLaughlin, 1996; Judge and Douglas, 1998; Sharma and Vredenburg, 1998; Melnyk et al., 2003; Sroufe, 2003). Among the first attempts to get empirical insight into this relationship is a study by Klassen and McLaughlin (1996). They find a significant relationship between positive environmental media attention (e.g. for a 'green investment') and increased stock returns and vice versa (i.e. negative media attention leading to decreased stock returns). The relationship remains statistically significant, after taking various other factors into account, such as firm size and type of media event (i.e. national versus international). In line with this, Russo and Fouts (1997) find a positive relationship between attention for environmental activities and financial performance in terms of return on assets (ROA). It appears also from their study that firms in a higher compared to lower growth industry, measured as the average increase in annual sales, perceive a stronger impact on ROA. The relationship remains significant after taking into account several control variables, including company size and growth rate.

Table 2.2. Quantitative studies on the relationship between environmental capabilities and firm performance.

Reference	Dependent variable(s)	Main results
Klassen and McLaughlin (1996)	Stock market performance	Positive media attention for the environmental performance of the firm is positively related to an increase in stock market return, and vice versa.
Russo and Fouts (1997)	Financial performance (ROA)	Environmental management performance is positively related to financial performance, especially in high-growth industries (measured as annual increase in sales).
Judge and Douglas (1998)	Financial and environmental performance	The integration of environmental issues in strategic planning is positively related to financial and environmental performance.
Aragón-Correa (1998)	Environmental performance	Business strategy pro-activity contributes positively to both traditionally regulated command-and-control as well as model pollution prevention environmental practices.
Klassen and Whybark (1999)	Manufacturing performance	Pollution prevention technologies contribute positively to manufacturing performance (except for production costs) and negatively to pollution control technologies.
Sharma and Vredenburg (1998)	Organizational capabilities and competitive benefits	Pro-active environmental strategies are associated with organizational capabilities, which create competitive benefits in terms of innovativeness.
Christmann (2000)	Cost advantage of environmental strategy	Capabilities related to prospector strategy are complementary to different environmental practices. Innovation in pollution prevention technologies delivers costs advantages.
Sroufe (2003)	Operations performance	Environmental management system implementation positively influences operations performance, directly and indirectly via capabilities on process (re)design, recycling, and waste practices.
Melnyk et al. (2003)	Operations performance and environmental options	ISO14001 certification is positively related to operations performance and different environmental management practices, like product/process (re)design, disassembly, and recycling.
González-Benito and González-Benito (2005)	Operational, marketing, and financial performance (ROA)	Both positive and negative relationships exist between different environmental management practices and operational as well as marketing performance, while no significant relationship with financial performance is found.

Description/comments	Data source
Positive environmental media attention includes environmental awards reported in the media. Financial event methodology has been applied.	Sample of 166 positive and negative U.S. environmental events reported in the media (NEXIS database)
Environmental performance is divided into less and more pro-active practices. All variables are obtained from a secondary data source. Regression analyses are carried out.	Sample of 243 U.S. firms from various industries (FRDC database)
Financial and environmental performance consisted of multiple items. Structural equation modeling is applied.	Survey of 196 U.S. firms ($\geq$ 20 employees) from multiple industries (response rate: 30%)
Strategic pro-activity is measured in concordance with the strategy typologies of Miles and Snow (e.g. prospector vs. reactor). Regression analyses are carried out.	Survey of 105 Spanish firms from multiple industries (response rate 53%)
Performance is divided into: costs, quality, speed (of delivery), and flexibility. Regression analyses are carried out.	Survey of 66 firms ($\geq$ 50 employees) from the U.S. furniture industry (response rate: 28%)
The used constructs comprised a high number of variables ($\alpha \geq 0.80$), e.g. 95 items used to operationalize environmental strategy. Regression analyses are carried out.	Survey of 99 firms with (annual revenues $\geq$ \$20 million) from Canadian gas and oil industry (response rate: 90%)
Respondents were asked to indicate different costs (e.g. of compliance with regulation) relative to their competitors ($\alpha \geq 0.70$). Regression analyses are applied.	Survey of 88 firms (annual sales $\geq$ \$100 million) from US chemical industry (response rate: 20%)
Environmental management performance was operationalized with 19 variables. Operations performance was broadly defined, including improved quality and reputation enhancement. Structural equation modeling was applied.	Survey of 1118 U.S. manufacturing firms from different types of industry (response rate: 10%)
A distinction is made between firms with and without a formal environmental management system, as well as firms with ISO14001 certification. Regression analyses are carried out.	Similar to Sroufe (2003)
The environmental management practices are: planning and organization, logistics processes, product design, and internal production management ($\alpha \geq 0.68$). Financial performance is collected from a secondary data source. Linear regression analyses are performed.	Survey of 186 Spanish firms (>100 employees) from different industries (response rate: 43%)

Table 2.2. Continued.

Reference	Dependent variable(s)	Main results
Menguc and Ozanne (2005)	Financial performance	Natural environmental orientation is positively related to market share and profit after tax, but negatively to sales growth.

Of interest in the present study is the research that takes a closer look at care for the environment through consideration of firm strategy and (other) organizational capabilities. Aragón-Correa (1998) was among the first to empirically test the relationship between strategic pro-activity and attention for the environment. He bases his findings on the survey results of 105 Spanish firms from different sectors (e.g. food and beverage, automotive, chemical, and retail). The strategic pro-activity is measured by Miles and Snow's (1978) strategy typologies (e.g. prospectors, defenders, and analyzers). A factor analysis is carried out on different environmental management activities. The result is used for a subsequent cluster analysis. The differences between the cluster groups indicate that non-compliance with environmental regulations can be associated with a significantly lower level of strategic pro-activity compared to the other environmental practices. Moreover, a regression analysis shows that strategic pro-activity explains the adoption of several environmental management capabilities. The author suggests that competitive advantages may be achieved by a positive trade-off between strategic pro-activity and the development of environmental management capabilities.

Building on the previous research, Christmann (2000) proposes that the cost advantages of taking care of the environment are moderated by so-called complementary capabilities that can be associated with a prospector strategy. Following Teece (1986), she defines these as *capabilities that allow firms to capture profits associated with a strategy, technology, or innovation.* The respondents are asked to indicate whether environmental activities imply cost advantages in comparison with competitors (e.g. fewer costs to conform to environmental regulation). Based on a factorial data reduction, three environmental practices are discerned: pollution prevention, environmental innovation (e.g. development of new environmentally friendly process technologies and products), and early timing of environmental attention (e.g. the first to use clean technologies). Various complementary capabilities are discerned relating to the attention for process innovation, such as being the first to introduce new production methods and technologies. A regression analysis shows that complementary capabilities contribute significantly to costs reductions gained in all three environmental practices. Interestingly, this finding implicates that the development of environmental management capabilities might be positively associated with a proactive firm strategy.

Description/comments	Data source
The natural environmental orientation is divided into: corporate social responsibility, environmental entrepreneurship, and commitment ($\alpha \geq 0.88$). Structural equation modeling is applied.	Survey of 140 Australian firms from different industries (response rate: 29%)

Sroufe (2003) investigates the affect of an environmental management system on operations performance through the development of various environmental organizational capabilities (see Table 2.2). Operations performance is broadly defined, including improved product quality, cost reductions, and enhanced reputation (among others). The considered environmental practices are environmentally friendly product (re)design, recycling activities, and attention for waste reduction. A structural equation model confirms that an environmental management system has, in a strictly statistical sense, not only a direct impact on business performance, but also an indirect positive impact on the development of the different environmental practices. A study by González-Benito and González-Benito (2005) indicates, though, that such positive effects may not always emerge. They find that only a limited number of environmental organizational capabilities are significantly related to different dimensions of business performance, such as operational performance (e.g. operational costs and flexibility), marketing, and financial performance. For example, a regression analysis supports a positive relationship between environmental logistics capabilities (e.g. use of clean transportation methods) and a lean operational performance in terms of quality and flexibility. However, a negative relationship is found between reduction of environmental emissions and mass operational performance (e.g. operational costs): this might be explained by costs made to implement environmental cleaning-up measures.

2.4 Concluding remarks

This chapter discussed environmental management as an essential element of corporate social responsibility (CSR), which can, in turn, be interpreted as an important step towards sustainable production. In this context, the development of environmental capabilities is essential. The present study can be discerned from previous research. First, it examines the joint impact of different stakeholders from the business network from a longitudinal perspective. A longitudinal evaluation is lacking in previous research, although some studies indicate changing stakeholder impacts in time. Second, it assesses the relationship with government in terms of different governmental strategies, including the use of public-private environmental agreements. This remains uninvestigated in the previous studies. Third, it considers the influence of firm strategy and enabling environmental capabilities (e.g. food quality and safety care systems) that are expected to be important in stimulating the development of environmental management capabilities. In doing so, it builds on previous studies which show that innovative business strategies can contribute to proactive environmental practices.

3. Study domain

This chapter discusses the study domain. Section 3.1 looks at the growing international attention for the environment reflected in various environmental initiatives, such as the Kyoto Protocol. It refers to public and private organizations that have been established to diffuse environmental awareness and to stimulate firms to implement environmentally responsible business practices. Furthermore, various standards and guidelines on environmental management are discussed. Section 3.2 focuses on the Dutch situation. It discusses characteristics of the governmental environmental policy, including the use of various public-private environmental voluntary agreements (or environmental covenants). Section 3.3 continues with economic and environmental characteristics of the Dutch food and beverage industry. Further, it elaborates on the attention paid to environmental issues by intermediary organizations (e.g. production boards and branch-organizations) in the Dutch food and beverage industry. Finally, Section 3.4 provides the concluding remarks with respect to this chapter.

3.1 Environmental attention, initiatives and standards

Incentives to take care of environmental risks and related pressures on the environment grew in response to a number of environmental incidents worldwide over the past few decades. Table 3.1 gives an overview of important industrial incidents, such as the release of toxic chemicals, oil spills, and the contamination of rivers and soils.

Table 3.1 shows that the 1980s in particular were dominated by several major incidents that triggered environmental awareness. The importance of taking care of the environment was not only fed by reasons to avoid environmental pressures, but also to protect corporate

Table 3.1. Environmental incidents with major societal consequences.

1978	Oil spill, Amoco Cadiz, France
1980s	Soil and river pollution scandals in many countries
1986	Rhine pollution, Sandoz factory, Switzerland
1989	Oil spill, Exxon Valdez, Alaska, USA
1991	Burning oil wells during Gulf War, Kuwait
1999	Oil spill tanker Erika, Total, Bretagne, France
2005	Explosion of a chemical factory, Harbin, China
2006	Dumping of chemical waste, Ivory Coast, Africa

Adapted from: Kolk (2000).

image[3]. The oil spillage on the coast of Alaska in 1989 (see Table 3.1) is said to still negatively affect the reputation of ExxonMobil (Van Tulder and Van der Zwart, 2006). At the national level, environmental incidents causing soil and river pollution during the 1980s fuelled debates on necessary changes of businesses towards environmentally friendly practices. In the Netherlands a major environmental scandal in 1983 included the discovery of houses built on a site where chemicals were dumped (Hajer, 1995). Environmental surveys among multinational firms during the 1990s show that the predominant environmental concern was to comply with environmental regulations and to avoid environmental incidents (McKinsey, 1991; UNCTAD, 1993).

The worldwide attention for the environment has resulted in many international environmental initiatives. Important developments in this context are the establishment of the United Nations Framework Convention on Climate Change (UNFCCC) in 1992 and the international agreement on the Kyoto Protocol in 1997. Both initiatives aim to reduce the release of greenhouse gases that cause global warming. The Kyoto Protocol came into force in February 2005, after Russia signed it. The participation of Russia was necessary to reach the condition of at least 55 participating countries from the list of so-called Annex-I Parties that were responsible for 55% of the greenhouse gas emissions of that group in 1990[4]. The European Union (EU) signed the protocol in 2002. The Netherlands is among the six European countries which are expected to be unable to reach their Kyoto targets in their own country (Albrecht and Arts, 2005). However, it can meet them through compensation for clean investments in Eastern Europe (called *joint implementation* in the Kyoto protocol) and developmental countries (called the *clean development mechanism*). During the last G8 summit in Germany it was agreed that the United Nations should play a key role in the establishment of a post-Kyoto Protocol after 2012[5]. Interestingly, the USA apparently also has a positive attitude towards such an initiative, even though it has still not accepted the Kyoto Protocol.

Parallel to the growing attention for the protection of the environment, industry and non-governmental organizations (NGOs, such as environmental organizations) have organized themselves (Brophy, 2004). One example is the Coalition of Environmentally Responsible

---

[3] Shell developed the Tripod-model (Dankaart 2007). Basically, the model assumes that imperfect decisions ('Latent failures') may cause increased risk of environmental incidents ('Preconditions') which could eventually lead to outbreaks ('Active failures'). If these outbreaks are not anticipated by so-called 'Last defense' measures (e.g. chemical liquid or fire-resistantce floors and building constructions), they might cause environmental incidents. A Tripod analysis typically asks why measures have been taken and relates incidents to management system failures rather than to technical failures.

[4] The Annex-I Parties participate in the UNFCCC and they consist of industrialized countries (like the EU, US, Canada, and Australia). They have together a large part in emissions of greenhouse gases worldwide. Other countries (such as China, India, and Brazil) have also access to the Kyoto Protocol: they are referred to as non-Annex-I Parties. In total, 171 countries have committed themselves to the protocol (date April 2007, source: http://unfcc.int).

[5] *Volkskrant* June 9 2007.

Economies (CERES) in the USA. It consists of both industry and NGOs, which have together launched the CERES Principles on Environmentally Responsible Management in 1989 (see: www.ceres.org). CERES was also responsible for the initiation of the Global Reporting Initiative (GRI) in 1997. The GRI is now an independent organization affiliated with corporations and NGOs all over the world and its guidelines on environmental reporting are well known (see: www.globalreporting.org). Another example is the World Business Council for Sustainable Development (WBCSD). It was established in 1991 and it currently involves 190 firms from 35 different countries. The WBCS tries to assist firms with turning their organization into an environmentally responsible and sustainable business (see: www.wbcsd.org).

Guidelines and standards have also emerged on the implementation of environmental management systems (Starkey, 2004). An environmental management system can be defined as the part of the overall management system which aims to reduce pollution. It facilitates the translation of environmental ambitions into environmental performance using the principle of continual improvement of quality in the Deming plan-do-check-act cycle (Netherwood, 2004). The ISO14001 standards on the implementation of an environmental management system are well known worldwide. Although firms can be certified based on 'self-certification', certification by an independent third party prevails. Important conditions for ISO14001 certification are compliance with environmental regulations, the proper functioning of all elements of the system and whether the organization indeed tries to reduce its environmental impact. Box 3.1 provides more insight into ISO14001.

The EU introduced the European Community's Eco-management and Auditing Scheme (EMAS) in 1995. The EMAS standards were originally directed at industrial production manufacturing sites only, but their scope has been extended to include different types of organizations (schools, offices, service companies, etc.). The standards induce requirements with respect to the implementation of an environmental management system, as well as on environmental reporting, which is referred to as the release of an *environmental statement*. The main difference between EMAS and ISO14001 is the mandatory environmental reporting. Germany accounts for the highest number of EMAS registered firms (1,487), because the German government has stimulated its implementation by means of regulatory relief and subsidies (Wätzold *et al.*, 2001). Spain (653) and Italy (529) are two other countries with a high number of EMAS certified organizations (date January 2007, see: ec.europa.eu/environment/emas).

Box 3.1. ISO14001.

ISO14001 defines an environmental management system as *the part of the overall management system that includes organizational structure, planning activities, responsibilities, practices, procedures, processes, and resources for developing, implementing, achieving, reviewing, and maintaining the environmental policy.* In short, it states that the system facilitates the implementation of environmental policy and, as such, can contribute to the reduction of the environmental impact of a firm. Based on the continuous improvement of quality in the Deming plan-do-check-act cycle, an environmental management system strives to reduce the pressure on the environment through the formulation and achievement of new policy targets (Martin, 1998; Kolk, 2000; Netherwood, 2004). According to ISO14001, the planning stage encompasses the formulation of an environmental policy including environmental targets that at least cover all legal requirements. An environmental action program should be formulated to implement environmental measures to achieve the policy targets. The doing stage refers to the division of environmental responsibilities and tasks over the different departments and people in the organization. It also includes environmental training of personnel and facilitation of documentation on environmental activities. The checking stage includes monitoring of environmental emissions by an environmental audit, in order to check whether the environmental policy targets have been achieved. Finally, the acting stage refers to processing the environmental information from the audit by means of a management review. The outcomes of this review should be used to decide on reformulation of environmental policy targets, in order to improve the environmental performance following the same cycle of continual improvement.

3.2 Dutch environmental policy

Two important characteristics of Dutch environmental policy are the adoption of participatory environmental policy arrangements and transferred environmental responsibilities from the central government to lower governmental bodies, in order to tackle environmental problems more efficiently (Mol *et al.*, 2000; Jordan and Liefferink, 2004). The latter refers to regional (e.g. provinces and water authorities) and local (e.g. municipalities) governmental bodies that take care of granting environmental permits and monitoring of environmental emissions (Mol *et al.*, 2001). Not all firms in the Netherlands are required to get a governmental environmental permit, since an exception is made for several types of small firms and/or firms with a limited environmental impact, such as butcheries and bakeries (Backes and Nelissen, 2000; Jurgens and Wiering, 2000). They have to act in concordance with generic environmental requirements on noise, smell, soil pollution, etc. The law prescribes whether firms have to submit their request for an environmental permit at the municipality or province (or at a governmental Ministry, in case of military organizations or nuclear power plants). Most firms have to deal with the municipality, because the province deals in particular firms with a high level of pollution (e.g. chemical industry). In addition to an environmental permit, there are other environmental regulatory obligations that have to be taken into account. Important examples are environmental legislation to protect the soil and water quality. Environmental permits are, for instance, required for waste water disposal, which is taken care of by the Dutch water boards. Hence, firms have to get different environmental permits from different governmental agencies.

Instruments used by the Dutch government to stimulate firms to take responsibility for care of the environment include the provision of environmental information to increase environmental awareness as well as positive incentives to clean up, like subsidies (Van der Kolk, 1987; Geelhoed, 1992). Positive governmental incentives are available to stimulate firms to implement an environmental management system (see Box 3.2). Important in this context is the granting of less detailed governmental environmental permits (Braakhuis et al., 1995: 88; Vermeulen, 2002: 77). Environmental permits are in general voluminous and complex documents, but the less detailed permits show an overview of the main lines only, as the details are left to the firm to deal with in its environmental management system. Furthermore, governmental environmental reporting was introduced in 1999 which made it mandatory for firms with a high level of pollution to report annually on their environmental emissions and clean-up measures (Bremmers, 2000). The reporting requirements may stimulate a firm to implement an environmental management system, because it can help with the provision of the necessary information for the report (Huizing, 1993).

The Netherlands was one of the first countries to extend its environmental policy from 'command-and-control' towards more participatory policy arrangements, including voluntary public-private environmental agreements or environmental covenants (Glasbergen and Driessen, 2002). Together with Germany, it accounts for the highest number of environmental covenants worldwide (Mol et al., 2000). One example is the Long-Term Agreement Energy, which aims to assist firms to reduce their energy consumption (see: www.senternovem.nl; with information available in English). Since its introduction in 1991, it has been followed up by a renewed agreement in 2001[6]. Furthermore, the Energy Efficiency Benchmarking Covenant is targeted at firms with a high level of energy consumption, such as the chemical industry, but also large breweries and sugar processors in the food and beverage industry. It was established in 1999 and aims to improve the energy efficiency of industry through the implementation of advanced and energy-efficient processing technologies (see: www.benchmarking-energie. nl; with information in English). Another example is the Packaging Covenant, which ended in January 2006. Its goal was to reduce the amount of packaging material used. It has been followed up by legislation based on EU environmental policy requirements (VROM, 2006b). It prescribes that individual firms have to pay for the return, processing and/or recycling of the packaging materials they use or import.

---

[6] It should be noted that individual firms are allowed to participate in this covenant providing that their branch-organization and/or product board (see also Section 3.3.3) has signed the agreement as well.

Box 3.2. Introduction of environmental management standards.

The Dutch employer organizations VNO and NCW jointly addressed a proposal in 1985 in which the government was asked to stimulate the implementation of environmental management systems. With this initiative, they tried to pre-empt an increase in governmental environmental regulation (Van der Kolk, 1987). Several pilot projects were set up in firms with governmental support (Van der Kolk, 1987; Braakhuis, Gijtenbeek and Hafkamp, 1995). These projects increased the firms' attention to implement an environmental management system, because financial savings could be achieved (e.g. less energy usage and waste production). Firms have never been obliged to implement an environmental management system. In 1989, though, the government did formulate two conditions to the voluntary implementation. First, firms with substantive environmental impact should have an environmental management system in place by 1995. Second, the minor polluting firms should also be encouraged to implement an environmental management system or parts of it, tailored to their situation, in all relevant sectors by 1995. An empirical investigation in 1997 showed that the second objective was achieved, but not the first one (RMK, 1997). However, it was concluded that most firms were preparing for the implementation of an environmental management system. The government therefore decided not to anchor the implementation of such a system in environmental regulation. Moreover, international attention for formal standardization (such as ISO14001 and EMAS) made the introduction of national standards less useful (Braakhuis, Gijtenbeek and Hafkamp, 1995). In total, 1,132 organizations (also other than firms) in the Netherlands are ISO14001-certified (by a third party certifier), of which 60 are food and beverage firms. Only 15 firms (among them Lyondell Chemical and NedCar) are EMAS-certified. None of them belong to the food and beverage industry (date January 2007, source: www.sccm.nl)

3.3 The Dutch food and beverage industry

3.3.1 Economic indicators[7]

The food and beverage industry is part of the agri-food sector of the Netherlands (see Box 3.3). It comprises about 4,500 firms providing employment to 144,000 people in total. In 2003 about 66% of them were employed in the 245 food and beverage firms with 100 or more employees (LEI, 2006). Table 3.1 shows the economic performance of the Dutch food and beverage industry compared to the Dutch industry as a whole.

Table 3.1 shows that the contribution of the Dutch food and beverage firms with 100 or more employees to net turnover, as part of total industry in the Netherlands, has slightly

---

[7] Information on the economic characteristics of the Dutch food and beverage industry, which are presented in this section, has been obtained mainly from the Dutch Agricultural Research Institute (see: www.lei.nl)

Table 3.1. Economic characteristics of the food and beverage industry compared to total Dutch industry (firms with 100 or more employees).

	2000	2001	2002	2003
Net turnover (x 10^9 Euros)	39.0	42.7	42.4	42.3
Contribution to the Dutch industry (% net turnover)	24.0	26.3	26.6	27.2
Export ratio (% net turnover)	47	43	41	46
Employment (x 1000 employees)	95.3	99.0	94.5	94.9
Contribution to Dutch industry (% employment)	18.8	19.5	19.2	20.1

Source: LEI (2003; 2004; 2005; 2006).

increased and stabilized at around 27%. The distribution of net-turnover over the sectors looked as follows in 2003: fat and oil (14%), meat (12%), mineral water and soda (9%), cocoa (7%), animal feed (6%), vegetables and fruit (6%), grain mill products (4%), bakery (4%), fish processing (1%), and others (38%), which includes the dairy industry for confidentiality reasons. It shows that the largest proportion of net turnover was generated by the fat and oil sector, followed by the meat sector and mineral water and soda. By contrast, the lowest stakes in net turnover can be attributed to bakery and fish processing. One reason for this might be the low added value generated by the fish processing firms and bakery factories. The export figures in Table 3.1 show that a substantive part of it is generated through export, namely 46% in that same year. Despite a slight decline in absolute numbers, the share of these firms in national industrial employment increased to 20% of the total employment in Dutch industry. In conclusion, the Dutch food and beverage industry represents a vital part of the economy in the Netherlands.

Box 3.3. The Dutch agri-food sector.

The Dutch agri-food sector consists of primary businesses (e.g. dairy and crop farms), the food and beverage industry, distributors (suppliers of logistical services), and a category 'other', consisting of suppliers of services, machines, and maintenance (LNV, 2005). The gross added value of the Dutch agri-food sector has hardly changed between 2001 and 2004. In line with this, the share in the national gross added value declined slightly. Employment also decreased, both in absolute terms, as well as compared to total employment in the Netherlands (from 11.1% in 2001 to 10.1% in 2004). This can be related to the decrease in farms. In particular, the number of small compared to large farms has declined rapidly over the past few years. An important reason for this is the increasing pressure on economic margins.

Table 3.2 shows a list of the largest Dutch food and beverage firms in 2005. It indicates that Unilever was by far the largest Dutch food-processing firm. However, it should be taken into account that it also produces products other than food, such as washing powder and personal care products. Notably, the total number of employees worldwide, including the Netherlands, American, as well as Asian countries, is larger than that of all other food and beverage firms with 100 or more employees taken together in the Netherlands. The same goes for turnover, of which about 60% was generated by food and beverage products. However, the turnover generated in the Netherlands is far less, see Table 3.1. Other large firms are Heineken, Vion, Friesland Foods, and Nutreco (see Table 3.2).

Table 3.2. List of the largest Dutch food and beverage firms in 2005.

Name	Total turnover worldwide (x 10^9 Euros)	Turnover Netherlands (x 10^9 Euros)	Total number of employees worldwide (x 1000)	Category of products
Unilever	39.7	1.2	206	Food products (among others)
Heineken	10.8	N.A.	64	Drinks
Vion	6.3	N.A.	14	Meat
Friesland Foods	4.4	1.4	16	Dairy
Nutreco[1]	3.9	1.0	13	Animal feed
Campina	3.6	1.1	7	Dairy
CSM	2.6	0.3	8	Sugar, bakery ingredients, etc.
Numico	2.0	N.A.	14	Baby and clinical food
Wessanen	1.9	0.2	7	Biological food and snacks
Provimi	1.6	N.A.	8	Animal feed
Cosun	1.3	0.5	4	Food products and ingredients

[1] 2004.
N.A. = Not available.
Source: LEI (2006).

3.3.2 Environmental emissions[8]

Dutilh and Blijswijk (2004) have identified the following four most important environmental emissions in the Dutch food and beverage industry:
- Water usage and energy consumption.
- Emissions to water and air.
- The use of packaging materials.
- Production of organic waste.

Water usage and energy consumption

Water can be used for washing organic products (such as potatoes and sugar beets), cleaning processing equipment and machinery, and to produce steam. Only 10% of the total amount of water is actually used as a product ingredient. Many food and beverage firms reduced their water usage per kg of produced volume during the 1990s, due to the introduction of management systems (such as Total Productive Maintenance), which aim to increase the production efficiency (Dutilh and Blijswijk, 2004). The most recent figures on water usage by the food and beverage industry, available at the CBS (StatLine), date back to 2001. Total Dutch industry accounted for 3,307 billion m^3 of water, of which 248 billion m^3 (= 7.5%) was used by the food and beverage industry.

Energy consumption can be related to different food processing activities. It is, for instance, needed for heating (frying potatoes, baking bread, etc.) and cooling (freezing and storage of vegetables, meat, etc.). The energy consumption by the Dutch food and beverage industry decreased from 2000 till 2005, both in absolute terms as well as relatively compared to the total industry in the Netherlands. The latter can be observed in Figure 3.1. Hence, the food and beverage industry has reduced its energy consumption more than the average Dutch industry. Looking at the sector level relatively high levels of energy consumption can be attributed to the dairy, bakery, and the animal feed sector (Duthilh and Blijswijk, 2004). This can be explained by their production processes, such as energy needed to bake bread and pasteurize milk[9].

Emission to water and air

In general, water in the food and beverage industry is not re-used, because of the risk of food contamination. Consequently, it represents a relatively large share of the emissions to water (including both releases to surface water and the sewerage system). It is common to

---

[8] Information on the environmental characteristics of the Dutch food and beverage industry, which are presented in this section, has been obtained mainly from the online database 'StatLine' (see: statline.cbs.nl).

[9] Looking at the agri-food sector as a whole, it can be noted that cultivation under glass (e.g. vegetables in greenhouses) has by far the largest share of energy consumption: it equaled 1.3 times the energy consumption of the total Dutch food and beverage industry in 2003.

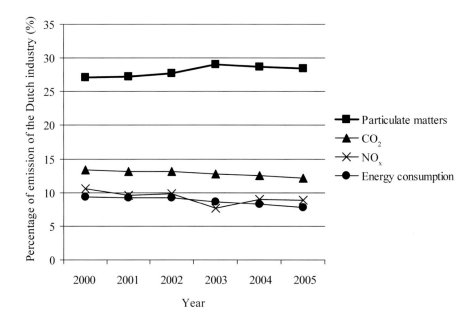

Figure 3.1. Share of the food and beverage industry in the total emissions of CO_2, NO_x, particulate matters, and energy consumption of the Dutch industry from 2000 to 2005.

express these emissions in terms of released equivalents of phosphate (P), which reflects the nutritive value. In 2000, 2003, and 2004, it was responsible for 1,495, 1,138, and 1,227 tons of phosphate emissions to water, respectively. It equaled 48%, 67%, and 72% of the total Dutch industry emissions in these years. It can therefore be concluded that its contribution to water emission of the total Dutch industry is substantive. The vegetables and fruit as well as the dairy sector account for a relatively high amount of water emissions in the Dutch food and beverage industry. This can be attributed to regular cleaning of the production process and washing of raw materials, such as harvested potatoes and sugar beets.

Emissions of CO_2 and NO_x form an important source of global warming and acid rain. In the food and beverage industry, both CO_2 and NO_x releases can be related to energy consumption, for which fossil fuels are used, including oil and gas. Figure 3.1 shows that the release of these emissions as part of the total Dutch industry has hardly changed over the past few years: for CO_2 production it stabilized at 13% between 2000 and 2005 and for NO_x emissions at around 9%. However, the emissions have slightly declined in absolute values.

Another relevant emission is the release of so-called particulate matters. They can cause severe health damage, because they consist of tiny little substances that can affect the lungs. This is one of the reasons why its reduction is an important goal for the Dutch government (VROM, 2005). In the food and beverage industry, examples of activities which contribute to the diffusion of particulate matters are transportation and use of raw materials (in bulk volumes), such as during (un)loading of ships, trucks, and silos with grain corns, meal, or soy beans. Figure 3.1 shows that the food and beverage industry accounted, with increasing volumes, for about 25-30% of the emissions of particulate matters caused by the total Dutch industry from 2000 till 2005. Lastly, the release of smell will be a particularly important environmental issue to food and beverage firms, which are located close to a community. It has a direct negative effect on the local environment. Smell can be caused during the production process itself (e.g. at bakeries, breweries, and potato processors), but also because of handling or processing organic waste at the manufacturing site.

The use of packaging material

The food and beverage industry accounted for 60% of the packaging material used by the total Dutch industry in 2001 (Dutilh and Blijswijk, 2004). Important reasons for this large share are conservation of the freshness of the products (e.g. cucumbers and tomatoes) and the fact that products are delivered to consumers. This has consequences for the volume of single product units, which are suited to the consumer market and therefore limited in size: small compared to large packaging units require more material. Different substances are used for packaging, including glass and metal (pots for carrying soup, vegetables, mayonnaise, etc.), as well as paper and plastic (boxes for carrying coffee, tea, vegetables, etc.). The food and beverage industry was responsible for 95% of glass, 60% of metal, 50% of paper, and 40% of polyester used for packaging in 2001. About 60% consisted of recycled materials, such as paper, plastic, and glass. Recent figures on the use of packaging materials are not available, because they are confidential (NVC, 2003). It is expected, though, that the use of absolute amounts of packaging material has declined, while the percentage of recycled substances has increased, based on the participation in the packaging covenant.

Production of organic waste

Food and beverage firms produce different solid waste flows, most of which are organic. In slaughterhouses, it consists of the organic material that remains after the animals have been slaughtered and processed (bones, skin, blood, etc.). Examples from the vegetables and fruit sector are potato processors, which have waste flows with potato peels and pieces, after cutting the potatoes into different shapes (e.g. to make chips). Other firms from this sector are processors of green vegetables, which produce organic waste consisting of vegetables that are not suitable for consumption (like rotten or broken vegetables). The food and beverage industry was responsible for slightly more than 50% of the solid waste production by the Dutch industry in the period 2000 till 2003. The vegetables and fruit firms counted for 47% of this waste production in 2003, followed by 12% for the meat, 8% for the beverages, and 2% for the dairy sector.

About 90% of the waste flows in the Dutch food and beverage industry were recycled in the period 2000 till 2003. Organic waste is therefore often referred to as a 'by-product' rather than waste (Dutilh and Blijswijk, 2004). The vegetables and fruit firms can, for instance, deliver their organic waste to animal feed producers, who use it as an ingredient. It was also common for slaughterhouses to sell their organic waste to animal feed producers. However, since January 2001, in response to the outbreak of BSE in Europe (also known as 'mad cow disease'), the EU has restricted this practice in order to reduce the risk of contaminated meat penetrating the consumer market. Furthermore, food and beverage firms are increasingly using their organic waste to generate bio-fuels, which can either be sold or used at the manufacturing site (SenterNovem, 2005).

3.3.3 Attention for environmental and food quality issues

Looking at the different Dutch food and beverage sectors, public-private intermediary organizations, such as branch-organizations and product boards, can influence the attention for the environment. Branch-organizations are organized and managed by firms themselves. Their aim is to support and act on behalf of their members in discussion with the government, other branch-organizations (e.g. from buyers and/or suppliers), and societal groups. Product boards are so-called statutory industrial organizations. In contrast to branch-organizations, they have legal power to carry out regulative measures on behalf of the Dutch government. Notably, product boards focus on the 'production column'. Therefore, they include different levels of the supply chain (e.g. primary producers, processors, and retail), while branch-organizations concentrate on the sector level. The FNLI (Federation of the Dutch Food and Beverage Industry) acts as an important 'umbrella organization' for the Dutch food and beverage industry (see: www.nfli.nl). Individual firms as well as branch-organizations are affiliated with it. It pays attention to environmental issues, mainly in relation to governmental environmental regulation. By contrast, DuVo (Foundation for a Sustainable Food Chain), an initiative by several Dutch food and beverage firms (Unilever, Cosun, Numico, Campina, etc.), pays attention to environmental issues from the perspective of sustainable production (see: www.duvo.nl, with information in English). Box 3.4 provides examples of branch-organizations and product boards that are active in the Dutch food and beverage industry.

It is interesting to evaluate the most important food quality and safety standards, because they sometimes pay attention to environmental issues as well. Based on EU requirements, Dutch food and beverage firms are obliged to implement an HACCP system (Hazard Analysis and Critical Control Points). It aims to systematically identify and implement control points in the production process to ensure food safety. The requirements attributed to an HACCP system are closely related to ISO9001, which is a management system that aims to fulfill quality demands, as well as enhance customer satisfaction based on continual quality improvement. Furthermore, the BRC standards (British Retail Consortium) are important to those food and beverage firms that deliver to retailers in the U.K. It requires the adoption of HACCP, documented in a quality management system (e.g. ISO9001), as well as managerial control of environmental and personnel issues (among others). Last, the Dutch Product Board Animal Feed has developed GMP standards (Good Manufacturing Practice). It aims to ensure the safety of animal feed, which is used in different food chains, including meat processing and egg production. It induces, among other things, the implementation of HACCP requirements and a quality system, like ISO9001.

Box 3.4. Attention for environmental management in different Dutch food and beverage sectors.

Meat sector
The Dutch Product Board for Livestock, Meat, and Eggs concentrates on livestock farmers (e.g. cattle, pigs, and poultry) as well as meat and egg processors (see: www.pve.nl; with information available in English). The attention for environmental issues is oriented towards the primary sector in particular (e.g. environmental emissions from manure handling). Different branch-organizations are active in the Dutch meat sector. The poultry meat processing firms are united in NEPLUVI (Association of Dutch Poultry Processing Industries, see: www.nepluvi.nl) and the producers of meat snacks in AKSV (General Cookwares and Snack Producers Association, see: www.aksv.nl). The latter has a commission that concentrates on food quality and environmental issues. The Long-Term Agreement Energy is signed by both AKSV (in November 2002) and NEPLUVI (in September 2002). Notably, other branch-organizations have also signed this voluntary agreement in September 2002: the KNS (Royal Dutch Butcher Association, September 2002), the VNB (Association of Dutch Bacon Manufacturers, September 2002), and the VNV (Association of Dutch Meatproducts Industry, September 2002).

Bakery sector
The NVB (Dutch Bakery Association) is the branch-organization for Dutch medium and large bakery factories (see: www.nvbweb.nl). One of its tasks is to inform its members about new environmental and packaging regulations. Furthermore, various privately organized collaborative initiatives can be found in the Dutch bakery sector. One example is Optifood, which is part of Bake Five (see: www.bakefive.nl). It purchases ingredients, packaging material, and various services on behalf of different bakery factories. Various conditions apply to participation in Optifood, like food quality and safety requirements. Interestingly, its mission comprises attention for sustainability issues, including attention for environmental regulation and social aspects, such as the rejection of child labour.

Vegetables and fruit sector
The VIGEF (Association of the Dutch Fruit and Vegetable Processing Industries) is the branch-organization for Dutch food firms that process vegetables and fruits (see: www.vigef.nl). An important focus of their program is finding opportunities to reduce energy consumption. Notably, the research projects are financed by the Dutch Product Board for Horticulture (www.tuinbouw. nl; also information in English available). The Product Board pays attention to other horticultural sectors as well, such as flowers, flower bulbs, and tree production. Both the VIGEF and the product board signed the Long-Term Agreement Energy in December 2001.

Animal feed sector
The Product Board Animal Feed acts as an intermediary between government and animal feed firms, as well as meat producers (see: www.pdv.nl; information can be obtained in English as well). Their yearly report shows that they focus pre-dominantly on food safety and quality issues, understandably because of the recent occurrence of major animal diseases, like foot and mouth disease. Furthermore, the large animal feed producers have united themselves in TrusQ (www.trusq.nl), and medium and small producers established SAFE FEED. These two organizations also focus primarily on food quality and safety issues.

Beverage sector
The Dutch beverage sector consists of several branch-organizations that serve different firms. The CBK (Central Brewery Office) is the branch-organization for the Dutch breweries (see: www.cbk.nl). Achieving sustainable production is one of their policy targets. It signed both the Long-Term Agreement Energy (suitable to large breweries) and the Energy Efficiency Benchmarking Covenant. The NFI (Dutch Soft Drinks Association) is the branch-organization for Dutch beverage firms that produce soft drinks, such as soda and mineral waters (see: www.frisdrank.nl). It pays attention to environmental activities, such as the reduction of energy and water usage. An environmental commission tries to increase the attention for environmental management through individual firm support, exchange of 'best-practice' experiences, and continual search for environmental improvement projects.

Dairy sector
The NZO (Dutch Dairy Association) is the branch-organization for the Dutch dairy industry (see: www.nzo.nl). One of their working groups is devoted to taking care of environmental issues. Notably, the Dutch Dairy Board has handed over most of its tasks directed at the dairy industry to the NZO (see: www.prodzuivel.nl; with information available in English). Furthermore, the Dutch cheese-melting firms are united in NEDSMELT (Dutch Association of Cheese Melters). NZO and NEDSMELT have jointly signed the Long-Term Agreement Dairy Industry. The backbone of this agreement is the so-called Integral Environmental Task. It comprises an overview of all pollution reduction targets, as prescribed by governmental environmental regulations. In order to achieve these targets, firms are obliged to implement an environmental business plan. Furthermore, the agreement implies mandatory participation in the Long-Term Agreement Energy (NZO/NEDSMELT, 2005).

Other sectors
The following firms, branch-organizations, and product boards have signed the Long-Term Agreement Energy: the branch-organization for Dutch grain-mill processors (NVM) in November 2004, the cocoa-processing firms collectively signed in July 2006, the Product Board for Margarine, Fats, and Oils in December 2001, the branch-organization for potato-processing firms (VAVI) in November 2004, and the branch-organization for the coffee and tea sector (VNKT) in December 2001. The Dutch sugar processors signed the Energy Efficiency Benchmarking Covenant.

3.4 Concluding remarks

From this chapter it can be concluded that a number of public and private initiatives, such as the Kyoto Protocol and the emergence of public-private coalitions like CERES, increased the attention for the environment. The Dutch government was among the first to extend its 'command-and-control' environmental policy towards the accomplishment of public-private voluntary arrangements, including the Long-Term Energy Agreement. In this political discourse, the food and beverage firms are challenged to demonstrate their responsibility for care of the environment. Various (inter)national environmental guidelines and standards aim to provide support to green business, like ISO14001. Furthermore, this chapter showed that the Dutch food and beverage industry represents not only a vital part of the Dutch economy, but is also responsible for a substantial level of environmental pollution. Branch-organizations and/or product boards aim to diffuse attention for environmental issues in the different food and beverage sectors of the Netherlands.

4. Theoretical review

This chapter presents the theoretical background. Section 4.1 discusses the outside-in and inside-out perspective on environmental management. The first refers to the effect of industry forces on the firm (e.g. business network pressures), while the latter refers to the competitiveness of the firm through the development of dynamic capabilities (e.g. clean production technologies). Section 4.2 concentrates on business network impacts. The environmental attention in individual firms will be affected by the attention for environmental issues in the business network: different types of network relationships are therefore discussed. Special attention is paid to the supply chain, which can be interpreted as a special type of network in which products and services are exchanged between buyers and suppliers. Furthermore, stakeholder typologies are evaluated, which refer to the importance and/or position of different stakeholders vis-à-vis the firm. Section 4.3 then proceeds with an evaluation of the competence perspective, including a discussion on resources and capabilities. It evaluates the natural resource-based view (natural-RBV), which states that attention for the environment can be accompanied by competitive organizational capabilities, such as those related to the implementation of pollution prevention technologies. Next, this section elaborates on environmental strategy typologies: it deals, in particular, with typologies that describe defensive and proactive strategies for environmental management. Section 4.4 provides the concluding remarks with respect to this chapter.

4.1 Two main perspectives on environmental management

Two main perspectives on environmental management are the outside-in and inside-out perspective (e.g. Kolk, 2000)[10]. The outside-in perspective is rooted in industrial organization theory, which states that business competitiveness and thus continuity depends on the ability of the firm to adapt to industry forces (Porter, 1980; Porter, 1981). Porter (1980) argues that five major industry forces shape competitiveness, including the threat of new entrants (e.g. new competitors that produce at lower costs), the bargaining power of buyers and suppliers (e.g. large buyer groups may exercise power to reduce purchasing prices), the threat of substitute products or services (e.g. cheaper substitutes will reduce consumer demand), and rivalry among existing firms (e.g. competition on low cost production). In reaction to these forces, firms can adopt the following strategies: overall cost leadership, differentiation, and focus (Porter, 1980). The overall cost leadership strategy deals with competitive forces by producing at the lowest possible costs, whereas the differentiation strategy tries to create something that is unique in terms of design or brand image, technology, and/or customer service. The focus strategy aims to serve a niche market, which could comprise consumer wishes for environmentally friendly produced products. Building on the outside-in perspective, the network theory aims to

---

[10] Kolk (2000) also distinguishes the inside-in perspective with which she refers to studies that investigate organizational barriers to environmental management improvement (e.g. a lack of organizational resources and capabilities). However, the present study considers this as part of the inside-out perspective.

explore the interrelatedness between a firm and different actors in its business network, such as buyers and suppliers, competitors, government, as well as various societal groups (Håkansson and Snehota, 1989; Powell, 1990; Omta *et al.*, 2001). In line with this, the present study evaluates the different stakeholder influences from the business network on the adoption of environmental management capabilities (see Section 4.2).

The inside-out perspective is rooted in the competence perspective, which relates the competitiveness of the firm to the development of valuable resources, competences, and capabilities (Barney, 1986b; a; Prahalad and Hamel, 1990; Teece *et al.*, 1997). Following Barney (2001), resources can be defined as *the tangible and intangible assets a firm uses to choose and implement its strategies*. Tangible assets are visible, such as machinery and production halls, while intangible assets are not visible, like experience and knowledge of employees. The resource-based view (RBV) evaluates the competitiveness of organizational resources (Penrose, 1959; Wernerfelt, 1984; Barney, 1986b; Rumelt, 1991). It states that it depends on the degree to which a single or a bundle of resources is valuable (i.e. contributing to the firm's efficiency and effectiveness) and rare (i.e. not widely held). In addition, the duration of the competitive advantage is associated with the extent to which the resources are vulnerable for imitation (i.e. copied by other firms), substitution (i.e. replaced by other resources with equivalent functions), and transformation (i.e. purchased in resource markets). The present study considers the inside-out perspective by means of organizational capabilities that can enhance the attention for environmental management capabilities (see Section 4.3).

4.2 Outside-in perspective on environmental management

4.2.1 Business network embeddedness

The firm's contacts in its network can influence the attention for the environment (Clarke, 1998; Clarke and Roome, 1999; Verheul, 1999). For example, branch-organizations can increase environmental awareness by providing environmental information on clean technologies and commitment to environmental covenants (see also Section 3.3.3). But other network actors on which a firm depends, like chain actors (e.g. buyers and suppliers), can also increase attention for the environment, if they impose environmental requirements on the firm. Håkansson (1982) was among the first to acknowledge that a firm depends on other actors in its business network (e.g. buyers, suppliers, and competitors) to achieve business performance. A network can be defined as a set of interdependent business relationships that evolve from the manager's personal contacts or from market-based relationships (Claro, 2003). Benefits of network contacts can be achieved, if they provide access to valuable resources, such as suppliers that deliver environmentally friendly ingredients. Despite the fact that some authors stress that networks are characterized by reciprocal, preferential and mutually supportive actions between firms (Powell, 1990), debates have also emerged on their constraints: an important contribution, in this context, is made by Granovetter (1992). He makes a distinction between relational and structural embeddedness. Relational embeddedness is the strength of the ties comprising the network, which can be measured in terms of quality and depth of the exchanged information. Granovetter (1973; 1985) argues that people involved in a network with strong ties tend to act according to the dominant views that are adhered by the network actors. They may be unaware or ignore opportunities to acquire valuable resources outside the network. It is therefore stated that weak rather than strong ties are beneficial to get access to new and valuable resources. Uzzi (1996; 1997) discusses the importance of strong and weak ties as well, using different terminology. He argues that strong ties (or what he calls *embedded ties*) are important to link the firm to the network in order to quickly extract information from it. At the same time, though, the network itself should comprise both strong and weak (or what he calls *arm's-length*) ties to prevent its insulation from new valuable resources. In practice a firm could, for instance, be unaware of a new environmentally friendly technology, because there is a lack of attention for it in its sector. However, contacts with firms outside its own sector may provide valuable information on their experience with this new technology.

Structural embeddedness refers to the architecture of the network in terms of redundancy (Granovetter, 1992). It can be visualized in a simplified triangular network, consisting of firm A, B, and C: if ties exist between firm A and B, as well as between A and C (e.g. for regular environmental information exchange or environmental cooperation), the establishment of contacts between firm B and C would form a redundant tie in the network. A sparse network implies that there are few redundant ties and, vice versa, a dense network includes many redundant ties. Two main, but opposite, views on the importance of redundant ties are addressed by Coleman and Burt (see: Scholten, 2006). Coleman's *closure argument* stresses the importance of redundant ties, because they facilitate trust and reduce opportunism (Coleman, 1988; 1990). This can be understood from the fact that actors in a dense network, in which all actors know each other, will be less likely to take advantage of one another. Moreover, a dense network will facilitate information exchange, because of sufficiently available communication channels as well as adequate insight into each other's information needs. By contrast, Burt (1992) discusses the advantages of non-redundant ties, which he refers to as *structural holes*. Acquiring a structural hole can imply access to valuable new resources, such as environmental knowledge, which are not available to other firms. Furthermore, acquiring information from non-redundant contacts can deliver timing advantages, such as the implementation of new environmental technology that is not yet used by competitors. Lastly, non-redundant contacts can facilitate the extension of network relationships through referrals, which can, in turn, contribute to access to new valuable resources (Burt, 1992).

Rowley (1997) adds an interesting dimension to structural embeddedness, namely the (network) centrality concept: it refers to *an individual actor's position in the network relative to others*. Based on Brass and Burkhardt (1993), a distinction can be made between the following types of centrality: degree, closeness, and betweenness. Degree centrality is the number of contacts the firm has with others in the network. Closeness centrality measures the ability of the firm to get directly in contact with other firms: a high closeness centrality implies direct rather than indirect access to other actors in the network. Betweenness-centrality is the extent to which the firm acts as an intermediary between (groups of) firms. It can be used as a proxy for the power of the firm to influence other actors, because the central firm represents the only or one of the few linkages between other (groups of) firms. Consequently, it is an exclusive opportunity to influence the exchanged information, including the contents and frequency (e.g. the opportunity to withhold critical information). Looking at the Dutch food and beverage industry, branch-organizations are expected to have a high betweenness-centrality, because they represent multiple firms towards the government and/or societal groups, such as environmental organizations. Furthermore, a large supplier or buyer (e.g. retailer or a chain captain) may also be characterized by a high betweenness-centrality. It implies that they can exert pressure to comply with environmental requirements based on withholding critical resources in case of non-compliance.

4.2.2 Chain and network actors

A chain can be interpreted as a special type of network. Christopher (1998) defines it as *a network of organizations involved in upstream and downstream linkages in different processes and activities that produce value in the form of products and services in the hand of the ultimate consumer*. The upstream and downstream linkages refer illustratively to a vertically organized sequence of firms (e.g. buyers and suppliers) involved in manufacturing of products (Lambert and Cooper, 2000). Lazzarini *et al.* (2001) have visualized this with their netchain concept, which can be defined as *a set of networks comprised of horizontal ties between firms within a particular industry or group, such that these networks (or layers) are sequentially based on vertical ties between firms in different layers*. Buyers and supplier firms, as well as consumers belong to the different layers. By contrast, competitors and branch-organizations (among others) can be found in the same layer. Figure 4.1 shows an example of a generic netchain.

The implications of the netchain concept can be illustrated for a focal firm, indicated by the black spot in Figure 4.1. It shows linkages with two suppliers and two buyers. It also has contact with two other chain actors in the same layer (e.g. competitors), as well as a non-chain actor (e.g. a branch-organization). Lambert and Cooper (2000) make a distinction between different types of process links of the focal firm with other network actors. Managed process links are important for safeguarding business continuity, such as the deliverance of environmentally friendly raw materials. The extent to which contacts have to be actively managed depends, for instance, on the resource (inter)dependency with other firms (see Box 4.1). Monitored process links are not critical to the firm, but are of interest, because they might (indirectly) influence the firm. For example, analyzing competitors may be important for anticipating green marketing strategies. Furthermore, non-managed process links can be discerned, comprising linkages in the netchain in which the firm is not (actively) involved, such as the linkages of its supplier with other suppliers in the same layer (see Figure 4.1). Lastly, non-member process links refer to linkages with non-chain actors, which can effectuate impact on business operations. Examples of important non-members are branch-organizations, either as important information sources or as providers of environmental guidance on the

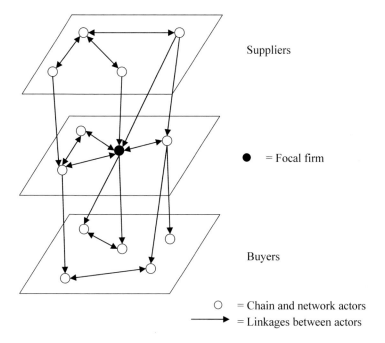

Suppliers

● = Focal firm

Buyers

○ = Chain and network actors
⟶ = Linkages between actors

Figure 4.1. A generic netchain (adapted from: Lazzarini et al., 2001).

implementation of environmental regulations and/or clean technologies. Furthermore, societal groups, like those formed by local communities and environmental organizations, are non-members too. Anticipation of their wishes is important to avoid adverse relationships and local actions that could hinder business operations.

Slack *et al.* (2001) define different relationships that can be discerned in a chain, using a continuum ranging from the spot market at one end and vertical integration at the other. A spot market is characterized by multiple buyers and suppliers, such as the flower auction in the Netherlands, while vertical integration refers to one firm covering the whole product chain, from harvesting the raw materials to manufacturing of consumer products. Although these extremes are not common in practice, different intermediate forms can be discerned, such as contracts, as well as coordinated profit sharing (e.g. licensing and franchising), alliances, and joint ventures (Slack *et al.*, 2001). In contrast to short-term contracts, which are often used for a single transaction, long-term contracts are typically used to cover multiple transactions, such as the delivery of raw materials on a daily or weekly basis. The benefits of long-term contracts can be found in increased certainty of delivery and less risk of opportunistic behavior based on an enduring relationship. Alliances refer to situations in which firms share valuable resources, including technologies, knowledge, employees, etc., while maintaining separated businesses with their own identity, culture, and structure. A joint venture can be interpreted as a special type of alliance, since it refers to the establishment of a new firm owned by the alliance partners. Referring forward to the results, different forms of environmental relationships with chain

---

Box 4.1. Resource (inter)dependency.

Stakeholders that possess resources, which are critical to the firm's main operations, will have a direct impact on the firm, while stakeholders that lack these resources can only exert an indirect influence (Pfeffer and Salancik, 1978; Frooman, 1999). Sharma and Henriques (2005) discuss four scenarios with respect to environmental resource (inter)dependency. First, if the stakeholders and the firm have high resource interdependency, they will try to exercise direct influence on the firm to use its resources such that their own objectives are achieved. An example is the establishment of an environmental label, as a result of cooperation between buyers and suppliers, in order to meet customer expectations. Second, if the stakeholders possess resources that are essential to the firm, while they are not dependent on it, they may try to directly influence the firm by threatening to withhold critical resources, if it does not come up to their requirements. Government can withdraw environmental permits, if a firm does not comply with environmental regulations. Third, if no resource dependency exists between the stakeholders and the firm, they are expected to try to influence the firm indirectly via other stakeholders. Local community and environmental organizations may, for instance, ask the government to force the firm to clean up. Fourth, if there is a high resource dependency of the stakeholders on the firm, but not vice versa, they will lack power to influence the firm. This could be the case for small firms, which have to deal with large counterparts, like buyers and suppliers.

---

actors were indicated by the interviewees, such as long-term contracts with suppliers (e.g. farms or other firms) to deliver environmentally friendly produced ingredients (see Section 6.2.2).

In the light of the previous discussion on network embeddedness, different relational contacts might be associated with typical network structures. For example, a large firm might want to cooperate with its suppliers on a franchise basis agreeing on fixed prices for the delivery of raw materials in exchange for service, information, and proprietary goods (e.g. harvesting machines and equipped storage places). Cooperation with buyers and/or suppliers might be beneficial to acquire valuable resources, such as environmental knowledge or contacts with other network actors. The reduction of transaction costs involved in upstream and downstream exchanges of products and services, such as costs for transportation and negotiations, can form a rationale for cooperation as well (Coase, 1937; Williamson, 1985). In general, transaction costs can be divided into three categories, namely costs for acquiring information (e.g. on prices and products), negotiations (e.g. writing contracts to stress delivery terms and liabilities), and monitoring or enforcement costs to exchange the negotiated goods and/or services (Hobbs, 1996). This is necessary not only to fulfill the negotiated agreements, but also to ensure that the activities are carried out as agreed in terms of quality and delivery conditions (e.g. frequency and volumes). A firm may, for instance, want to ensure that the supplied ingredients are produced on an environmentally friendly basis or transported under such conditions that product quality is guaranteed.

4.2.3 Stakeholder typologies

Firms are confronted with different stakeholder interests, including employees that want the firm to provide employment or local inhabitants that want it to take care of the local environment. Freeman (1984) defines a stakeholder as *any group or individual who can affect or is affected by the achievement of the organization's objective.* The business-ethical and instrumental perspective are the two main views on the stakeholder concept (Donaldson and Preston, 1995; Driscoll and Starik, 2004). The first is prescriptive and it stresses the moral obligation of firms to take care of stakeholder wishes. Broad and inconclusive definitions are used, which sometimes even allow for the inclusion of the environment as a stakeholder (e.g. Driscoll and Starik, 2004). By contrast, the instrumental perspective applies a narrower definition. It is used to *identify the connections, or lack of connections, between stakeholder management and the achievement of traditional corporate objectives* (Donaldson and Preston, 1995). Clarkson (1995) states, for instance, that stakeholders are *persons or groups that have, or claim, ownership, rights, or interests in a corporation and its activities, past, present, or future.* It comprises studies that try to determine stakeholder influences, using different stakeholder attributes, such as power and legitimacy (Mitchell *et al.*, 1997). The present study focuses on the instrumental perspective, because it aims to get more insight into the actual stakeholder influences rather than normative reasons for taking care of stakeholder wishes.

Stakeholders can be classified with respect to their position vis-à-vis the firm in internal and external stakeholders (Freeman, 1984). Henriques and Sadorsky (1999) refer to shareholders, management, and employees as internal stakeholders and regulators, public/community, and contractors/suppliers as external stakeholders. Clarkson (1995) uses a classification that is based on the importance of different stakeholder interests for the survival of the firm. Primary stakeholders are essential to survival, while secondary stakeholders are not. If the firm ignores the interests of primary stakeholders, such as supply chain partners, customers, and employees, it implies a severe threat for its continuity. However, secondary stakeholders, such as the local community and environmental organizations, can also affect the firm. A good relationship with them is important to maintain a social *license-to-produce* (Kagan *et al.*, 2003). Government occupies a special position: it is responsible for granting environmental permits, which force compliance with environmental regulations. It can also facilitate attention for the environment by means of subsidies and other supportive activities, like the provision of information on clean technologies and implementation guidelines. Notably, the different classifications of stakeholder groups can be combined. Buysse and Verbeke (2003) make, for instance, a distinction between primary stakeholders that are internal (e.g. employees) and external (e.g. buyers and suppliers) to the firm. Furthermore, Madsen and Ulhøi (2001) found that primary and secondary stakeholder groups could be further divided into those having a direct (e.g. government, owners, employees) versus an indirect (e.g. financial institutions) impact on the firm. In general, grouping of different stakeholders should be carried out carefully, because stakeholders classified in one group may still exercise different types of impact (McVea and Freeman, 2005).

Interestingly, Mitchell *et al.* (1997) defined typologies with respect to stakeholder salience, which is the degree to which managers give priority to competing stakeholder claims. A distinction is made between the following three stakeholder attributes: power, legitimacy, and urgency. Power is *the probability that one actor within a social relationship would be in a position to carry out his own will despite resistance* (Weber, 1947). Legitimacy can be defined as *a generalized perception or assumption that the actions of an entity are desirable, proper, or appropriate within some socially constructed system of norms, values, beliefs, and definitions* (Suchman, 1995). Claims made by legitimate stakeholders are perceived as trustworthy (Oliver, 1991; Suchman, 1995). Firms can perceive stakeholder wishes legitimate for pragmatic, moral, and cognitive reasons (Suchman, 1995). Self-interest and opportunism is at the basis of pragmatic legitimacy. It includes a situation in which stakeholder claims are perceived as legitimate in order to stay in business (e.g. the use of cheap, but environmentally unfriendly raw materials). Moral legitimacy is driven by ethical norms, for example if top management feels a moral obligation to take care of the environment. Furthermore, cognitive legitimacy refers to a special situation in which the expectations of a particular stakeholder are *taken-for-granted* and therefore considered without resistance (e.g. a lifetime buyer). The last stakeholder attribute, urgency, can be defined as *the degree to which stakeholder claims call for immediate attention* (Mitchell *et al.*, 1997). For example, changes in environmental regulations or environmentally driven crises (e.g. a chemical spillage may require an instant response from the firms involved.

Based on the three stakeholder attributes, Mitchell *et al.* (1997) make a distinction between stakeholder groups which are labeled according to how important it is to take care of them. The latent stakeholder group fosters expectations that are characterized by a high level of only one stakeholder attribute and, consequently, their wishes will be perceived as of minor importance. It comprises stakeholders with either a high level of power (e.g. suppliers that possess critical resources), legitimacy (e.g. local community to maintain a social license-to-produce), or urgency (e.g. environmental organizations that want the firm to reduce environmental pressure). The expectant group has demands, which can be characterized by two of the three attributes and their wishes will therefore be perceived as more important than the latent stakeholder claims. Lastly, the definitive stakeholders have the highest level of stakeholder salience, since they have the power to influence the firm with legitimate and urgent claims. The perceived importance of taking care of stakeholders is considered in the present study by measuring their overall impact on environmental management (see Chapter 5). In addition, special attention is paid to elements of the relationship with government which could have an important effect on their managerial salience (e.g. public-private environmental voluntary agreements and the importance of government as an environmental information source).

4.3 Inside-out perspective on environmental management

4.3.1 Resources, capabilities and competences

As already discussed in Section 4.1, the competence perspective evaluates competitiveness of firms through a consideration of their resources and capabilities. It is a fairly new conceptual perspective on the functioning of the firm that emerged during the 1990s. It brings together key ideas of the resource-based view (RBV), (dynamic) capabilities as well as core competence perspectives (Sanchez, 2004). Before discussing them in more detail, Table 4.1 provides an overview of the key assumptions and premises of the competence perspective in general as well as the resource-based view (RBV) and the (dynamic) capabilities framework.

Table 4.1. Overview of the basic assumptions of the competence perspective, resource-based view (RBV), and the dynamic capabilities framework.

Competence perspective
- Firms have certain "core" competences that span products and businesses, change more slowly than products, and arise from collective learning. Firms compete and achieve competitive advantage through creating and using their core competences;
- Knowledge resources are key sources of competitive advantage. A firm's strategic architecture influences its use of resources;
- Applying knowledge in action and learning are the foundations of a firm's competences and capabilities;
- Firms function as open systems of resource flows motivated by managers' perceptions of the strategic gaps a firm must close to achieve an acceptable level of goal attainment. Firms have distinctive strategic goals that lead to unique patterns of resource flows and competence building and leveraging activities;
- Competence leveraging drives short-term competitive dynamics, while competence building drives long-term competitive dynamics;
- The complexity and uncertainty inherent in managing resource flows in a dynamic environment make the "contest between managerial cognitions" in devising strategic logics a primary feature of competence-based competition;
- Firms rely on the use of both firm-specific and firm-addressable resources, and competition occurs in markets for key resources as well as in markets for products;
- Competence-based competition includes forms of cooperation (as well as competition) with providers of key resources;
- Firms' differing abilities in coordinating resources and resource flows and in managing their systematic interdependencies greatly influence competitive outcomes in dynamic environments;
- Creating a systematic organizational capacity for strategic flexibility may be the dominant logic for competence-based strategic management in dynamic environments.

Table 4.1. Continued.

Resource-Based View (RBV)
- Firm growth is motivated by the availability of the firm's resources;
- Firm growth is limited by management's recognition of productive opportunities suited to the firm's available resources. The ability to combine existing and new resources and willingness to accept the risk of using new resource combinations will allow a firm to meet new market demand;
- Resource position barriers can be created when experience in using resources lowers costs for incumbents and imposes higher costs on imitators;
- Diversification is an attempt to extend a firm's resource position barriers into new markets by combining its current resources with new resources;
- Mergers and acquisitions are attempts to acquire groups of attractive resources;
- Firms cannot create a sustained competitive advantage in markets with homogeneous and perfectly mobile resources;
- Creating a sustained competitive advantage depends on control of a firm resource endowment that includes resources that are heterogeneous, imperfectly mobile, valuable, rare, imperfectly imitable and non-substitutable;
- The rent-earning potential of resources results from the properties of resources that create asset mass efficiencies, asset mass interconnectedness and time compression diseconomies in firms' efforts to accumulate and to create assets.

Dynamic capabilities framework
- Changes in economic activities result from the learning and embedding of new skills in new organizational routines;
- Skill development in organizations follows natural trajectories determined by the organization's existing skill base and routines;
- Competitive advantage arises from a firm's current distinctive ways of coordinating and combining its difficult-to-trade and complementary assets;
- At any point in time, certain assets will be important determinants of a firm's ability to earn rents in a given market, i.e. they will be strategic industry factors, but these assets will be imperfectly predictable and subject to market failure;
- Managers' cognitive and social processes will determine the assets a firm acquires and with this its potential for generating organizational rents.

Adapted from: Sanchez (2001).

Resources are all assets owned by the firm, which can, according to the RBV, render competitive benefits if they are valuable and rare (Barney, 2001). The duration of the competitiveness depends on the extent to which they can be easily imitated, substituted by equally valuable resources, or transformed and sold in resource markets (as already discussed in Section 4.1). Examples of competitive resources are technical and marketing experience to produce at low costs or to deliver better product quality than competitors. Some authors argue that the RBV lacks attention for the dynamic effects on competitiveness associated with market developments (Priem and Butler, 2001a; b; Aragón-Correa and Sharma, 2003). After all, changing environmental consumer interests or new environmental regulations can negatively influence the competitiveness of the firm, if it fails to keep up with the new demands. In reaction to this, the contingent-RBV considers the effect of market components, such as the number of competitors and customer requirements, to determine resource competitiveness (Aragón-Correa and Sharma, 2003). Contingency scholars argue that business performance depends on fit, congruence, or alignment of what could be called the 'internal' and 'external' business environment (Duncan, 1972; Hofer, 1975; Venkatraman, 1984; 1989). Basically, the internal business environment refers to the available organizational resources, whereas the external business environment refers to market characteristics. Contingency scholars employ a system view: firms comprise of different organizational and technical elements, which have to function in alignment with each other and the business environment to enhance business performance (De Leeuw, 2000). Finding a profitable balance between the two is a dynamic rather than a static process, especially under fast-changing market conditions (Zajac *et al.*, 2000; Fredericks, 2005).

Theories on core competences and dynamic capabilities relate the development of competitive resources to the ability of the firm to use its resources in a way that it contributes to business performance. The two are closely related and some authors use the concepts interchangeably (e.g. Day, 1994). Also in the present study, competences and capabilities are used interchangeably (as already mentioned in Chapter 1). Prahalad and Hamel (1990) define core competences as *a complex harmonization of individual technology and production skills*. They enable the production of different products, which share the same competences, against reduced costs. Honda has, for instance, competences in building engines and power trains, giving it a distinctive competitive advantage not only in car building, but also in the production of motorcycles, lawn mowers, and generators (Prahalad and Hamel, 1990). The development of competences requires time to accumulate necessary knowledge and experience and to establish (new) working routines (Dierickx and Cool, 1989; Markides and Williamson, 1994; 1996). Capabilities are often associated with the capacity of a firm to adequately fulfill (certain) customers' demands and purchase markets in line with business strategy, which can be directed at serving a specific market or customer group (Amit and Schoemaker, 1993). In addition, dynamic capabilities are necessary to stay competitive in rapidly changing market environments (Teece *et al.*, 1997; Eisenhardt and Martin, 2000). It refers to *the firm's ability to integrate, build, and reconfigure internal and external competences to address rapidly changing environments* (Teece *et al.*, 1997). Dynamic capabilities are embedded in organizational routines, which makes their development path-dependent and hard to copy by competitors (Nelson and Winter, 1982; Teece *et al.*, 1997). Knowledge is a significant carrier of capabilities (Leonard-Barton, 1992; Adner and Helfat, 2003; Subramaniam and Youndt, 2005). Leonard-Barton (1992) makes a distinction between the following four knowledge dimensions of capabilities: (1) employee knowledge and skills, (2) technical systems capturing information and procedures (e.g. an environmental database), (3) managerial systems that aim to control and coordinate knowledge (e.g. an environmental management system), and (4) values and norms of the organization. It can be noted that the dimensions are inter-related: a managerial system may, for instance, contribute to knowledge of employees through training and educational programs.

4.3.2 Natural resource-based view

Hart (1995) is the founding author of the natural resource-based view (RBV). In the light of the ever-growing pressure on the environment, he argues that business continuity will increasingly depend on the development of capabilities to reduce the environmental impact. He regards capabilities as a bundle of resources being brought together to carry out particular value-added tasks, including designs for manufacturing and just-in-time production. Environmental capabilities are related to remediation of environmental impacts through the development of new working routines and the implementation of clean technologies. Furthermore, it comprises activities, such as environmentally responsible product (re)design and attention for life-cycle assessment (LCA). LCA can be defined as the evaluation and reduction of the environmental emissions that can be associated with products from *the extraction and use of raw materials through the eventual disposal of the components of the product and their decomposition back to the elements* (Welford, 2004). It aims to reduce the negative environmental effects of products from *cradle-to-grave*, or from *cradle-to-cradle* if waste outflows in one life-cycle can be used for another (McDonough and Braungart, 2002). LCA contributes to closed-loop production with respect to the avoidance of unnecessary environmental emissions.

The competitive advantages of environmental capabilities are analyzed, using the criteria of the RBV. In this context, Hart (1995) distinguishes between pollution prevention, product stewardship, and sustainable development capabilities. They are discussed below. Pollution prevention capabilities focus on reduction of environmental emissions at the level of the manufacturing site. Preventive measures are used to clean up, which can be integrated in the production process, by way of environmentally friendly production technologies and working routines (Lober, 1998). Competitive advantages can be achieved through the accumulation of environmental knowledge and experience, as well as improved efficiency and effectiveness of production processes resulting in less energy consumption and waste disposal. Product stewardship capabilities aim to reduce environmental emissions from the perspective of LCA, including the use of environmentally friendly produced raw materials, recycled product components and packaging (e.g. glass or paper), and environmentally responsible product design. Again, competitive advantages can be achieved in terms of accumulated knowledge and experience, which is essential for business performance. The sustainable development capabilities refer to a major turnaround in business processes to the benefit of the environment, using only environmentally friendly resources and production techniques. Hart (1995) argues that it requires a vision of the future that is shared by the firm and its stakeholders (buyers, suppliers, consumers, etc.) in order to overcome short-term disadvantages of sustainable production, such as high investment costs and increased purchasing prices. Sustainable development capabilities will make firms less dependent on depleting natural resources and contribute to competitiveness through the development of valuable knowledge (Hart, 1995; 1997).

Hart (1995) states that the competitiveness of pollution prevention, product stewardship, and sustainable development capabilities can be related to gaining social legitimacy. According to the natural-RBV, stakeholder impacts should be taken care of to maintain legitimacy and, moreover, to gain long-term competitive advantage. Pollution prevention capabilities enable the firm to show environmental excellence to stakeholders from the business network, such as societal groups, which will contribute to the social *license-to-produce* (Litz, 1996; Kagan *et al.*, 2003). Firms with product stewardship capabilities can go one step further by involving different stakeholders, such as environmental groups and consumers, in environmentally responsible product (re)design processes. This will contribute to the acceptance of their products and, in turn, it is expected to improve sales. Lastly, stakeholder involvement is already at the basis of the development of sustainable capabilities, since they are rooted in sharing resources (knowledge, technology, money, etc.) with different stakeholders (e.g. buyers, suppliers, and consumers) in order to find new and sustainable production technologies and marketing strategies.

4.3.3 Environmental strategy typologies

Strategy can be defined as *a series of goal-oriented decisions and actions that match an organization's skills and resources with the opportunities and threats in its [business] environment, to meet the need of markets and to fulfill stakeholder expectations* (Omta and Folstar, 2005). Likewise, environmental strategy can be defined as the part of firm strategy that aims to achieve and maintain a match of organizational resources and capabilities with environmental expectations addressed by stakeholders. Environmental strategy typologies have emerged to analyze environmental management practices. At the basis of the mainstream lies the reactive-defensive-accommodative-proactive (RDAP) scale, which is used by Caroll (1979) to evaluate corporate social responsibility (CSR). Clarkson (1991; 1995) refers to it in order to depict differences in social responsiveness.

Table 4.2 shows the characteristics of the different categories on the RDAP scale. Strategy in this context refers to the orientation towards social responsiveness. It moves from a reactive to a proactive attitude, which corresponds with denying responsibilities (doing less than legally or otherwise required) to anticipation (doing more than legally or otherwise required). Reasons for doing less than required can be a lack of resources (e.g. clean technologies) or willingness (e.g. other managerial priorities) to meet the demands. Doing more than required can be motivated by expected economic benefits, such as influencing (future) stakeholder expectations to the benefit of the firm (e.g. customer interest in environmentally responsible manufactured products). Although the RDAP scale is widely used to underpin environmental strategies, it provides a rather generic classification. Figure 4.2 relates it therefore to other

Table 4.2. The reactive-defensive-accommodative-proactive (RDAP) scale.

	Strategy	**Performance**
Reactive	Deny responsibility	Doing less than required
Defensive	Admit responsibility but fight it	Doing the least that is required
Accommodative	Accept responsibility	Doing all that is required
Proactive	Anticipate responsibility	Doing more than is required

Source: Clarkson (1995)

proposed environmental strategy typologies, such as defined by Roome (1994). He makes a distinction between non-compliance, compliance, compliance-plus, and environmental excellence (see Figure 4.2). Non-compliance refers to violation of environmental regulations, which may occur unintentionally, because of unawareness or time needed to comply with new regulatory demands. However, a firm can also deliberately neglect legal environmental liabilities, if the necessary measures are costly. Compliance with regulations implies acting in concordance with regulatory demands, but on a rather defensive basis. Roome (1994) even states that a compliance strategy implies a reactive attitude towards environmental care, because legal requirements represent often only part of the stakeholder pressures to clean up. Hence, a firm may act indifferently towards the environmental expectations of other stakeholders, such as the local community and environmental organizations, as long as it does not affect its social license-to-produce. The compliance-oriented firms use environmental technologies to control pollution rather than to prevent it. The compliance-plus category refers to firms that do more than required from the regulatory point of view. They constantly try to improve their environmental performance by integrating preventive environmental measures in their business operations. Lastly, environmental excellence is the situation in which top-management

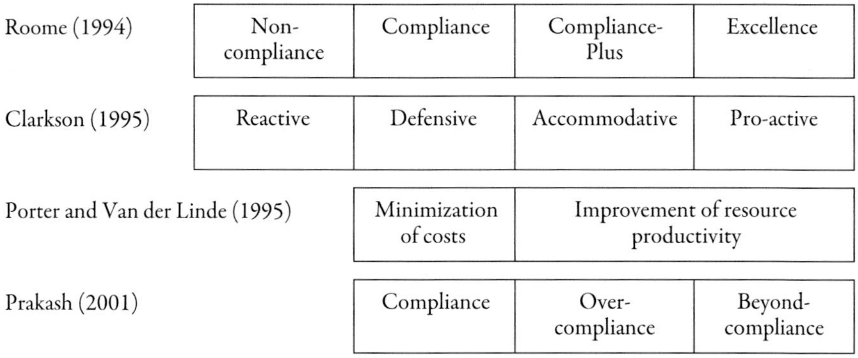

Roome (1994)	Non-compliance	Compliance	Compliance-Plus	Excellence
Clarkson (1995)	Reactive	Defensive	Accommodative	Pro-active
Porter and Van der Linde (1995)	Minimization of costs	Improvement of resource productivity		
Prakash (2001)	Compliance	Over-compliance	Beyond-compliance	

Figure 4.2. Different environmental strategy typologies.

shows a strong commitment to alter business for the benefit of the environment, covering all organizational functions (including production, R&D, marketing, etc.). In short, it implies that these firms try to gain commercial benefits from paying attention to the environment.

Porter and Van der Linde (1995) distinguishes between firms that aim to reduce the costs of compliance as much as possible, and those that constantly try to improve their resource productivity in terms of increasing production efficiency and effectiveness, as well as developing better products (see Figure 4.2). The cost reduction strategy is defensive to the point that it tries to keep the financial consequences of pollution as low as possible. This can be done through conversion of the pollution into something that is of value (e.g. re-cycling of waste) or into a substance that is less expensive to dispose of (e.g. less toxic). This strategy is probably adhered to by authors, who argue that it is costly to comply to environmental regulations, especially after the so-called *low-hanging-fruit* of inefficient production has been harvested (Walley and Whitehead, 1994; Newton and Harte, 1997). The firms that try to improve resource productivity acknowledge that taking care of the environment can render benefits, not only in terms of the previously discussed competitive organizational capabilities, but (consequently) also through meeting customer environmental wishes (Polonsky, 1995; Prakash, 2002; Peattie and Crane, 2005). Within this category, Prakash (2001) distinguishes between over- and beyond-compliance (see Figure 4.2). The first refers to firms that do more than legally required in line with, or as an extension of governmental environmental regulations, while the latter involves firms that also implement other environmental strategies, for example, satisfying the environmental wishes of their customers.

4.4 Concluding remarks

This chapter evaluated business network impacts and the importance of organizational resources, competences, and capabilities for business performance. It showed that perceived business network pressures depend on the power and legitimacy of the stakeholders involved, as well as the urgency of their claims. Their power increases when they have resources that are critical to the firm (see Box 4.1). It discussed the significance of network relationships for obtaining valuable environmental information and resources from other network actors. The elaboration of the natural-RBV showed that environmental capabilities, as needed for pollution prevention technologies and environmentally responsible product (re)design, can render competitive benefits through, for instance, improved production efficiency and effectiveness and meeting 'green' customer wishes. The evaluated environmental strategy typologies indicated that firms can adopt different strategies towards the environment: some focus on compliance with environmental regulation only, while others pay attention to the environment to differentiate themselves from competitors by environmentally responsible manufacturing.

From this chapter conclusions can also be drawn with respect to the interdependency between business network influences, firm strategy, and enabling environmental capabilities. Firms that aim for environmental excellence are expected to respond proactively to the wishes of the government and other stakeholders, such as environmental organizations and the local community. In line with previous empirical studies discussed in Chapter 2 (see Section 2.3 in particular), it is also assumed that environmental excellence can be associated with the development of valuable organizational capabilities, such as pollution prevention, which will contribute to business performance. By contrast, firms that adopt a defensive attitude towards environmental issues will not invest in the development of environmental capabilities, if not forced by governmental environmental regulations. The next chapter elaborates on these assumptions by presenting the conceptual model and propositions (see Section 5.2 and 5.4).

5. Study design

Section 5.1 elaborates on the employed mixed methodology using semi-structured interviews and two large-scale survey questionnaires. The questionnaires focus on the business network impact on the adoption of environmental management capabilities. The business network includes the government, chain and network actors (buyers and suppliers, competitors, branch-organizations, etc.), as well as societal groups. In addition, the survey of 2002 pays attention to the governmental relationships, while the survey of 2005 focuses on the importance of firm strategy and (other) organizational capabilities that may affect attention for the environment. The semi-structured interviews were carried out to get in-depth insight into the strategies and environmental activities of firms to respond to various business network influences. Section 5.2 presents the conceptual model. It considers the impact of the business network and firm characteristics on the adoption of environmental management capabilities. Section 5.3 elaborates on the operationalization of the conceptual model into research variables (or constructs) that are measured. The operational definitions of the used constructs are presented, as well as the unit of analysis, namely plant or business unit (e.g. in case of small firms). Section 5.4 presents the propositions, discussing the expected relationships in the conceptual model. Section 5.5 describes the procedures used for data gathering. The Chamber of Commerce has been consulted to obtain addresses of Dutch food and beverage firms. Section 5.6 evaluates the statistical techniques applied. It elaborates on the data analyses of both surveys, as well as the longitudinal analysis carried out to compare the results of 2002 to 2005 with respect to the availability of environmental management capabilities and business network influences. Section 5.7 provides the concluding remarks.

5.1 Overall research design

This study employed a mixed research design, including a qualitative (i.e. semi-structured interviews) and a quantitative (i.e. questionnaire surveys) research approach. Traditionally, they each reflect a distinct philosophy on the measurement of social behavior (Sechrest and Sidani, 1995; Onwuegbuzie, 2002; Onwuegbuzie and Daniel, 2003). At one extreme, qualitative purists argue that behavior can not be measured objectively, because it is indistinguishable from the context in which it is observed (e.g. the firm or sector). This would imply restrictions on the generalization of obtained results (Onwuegbuzie, 2002; Raskin, 2002). At the other extreme, quantitative purists state that behavior can be measured objectively and determined by universally valid predictors (Gerrard, 1990; Onwuegbuzie and Daniel, 2003). The caveat of this approach is a lack of attention for situational factors, such as firm characteristics, which may indeed limit the generalization of the results. Over the past few decades, there has been growing interest in methodological triangulation of qualitative and quantitative research approaches (Kaplan and Duchon, 1988; Morse, 1991; Sechrest and Sidani, 1995; Foss and Ellefsen, 2002). Sieber (1973) addresses examples of triangulation advantages for the quality of the study design, data collection, and analyses. They are in concordance with the experiences obtained during the present study. For instance, the interviews gave insight into environmental management practices and perceptions held by food and beverage firms, which contributed to the focus of the study and the development of the conceptual framework. It was useful for the construction of reliable and valid research variables and questionnaires, as well as the interpretation of the quantitative results. Overall, the results of the statistical analyses could be underpinned with findings of the in-depth interviews (see Chapter 6 and 7).

A distinction can be made between simultaneous and sequential methodological triangulation (Morse, 1991; Sechrest and Sidani, 1995). Simultaneous triangulation implies that two research approaches are executed in parallel, while sequential triangulation means that they are carried out one after another. The latter allows for a higher level of interaction between the approaches, which contributes to the benefits of triangulation (Morse, 1991). The present study can be typified as sequential triangulation of several semi-structured interviews and two survey questionnaires, which were conducted in 2002 and 2005. The interviews were carried out before and after the second survey. They were used to get more insight into environmental management practices as well as the reasons and motives that could explain the quantitative results. The aim of the survey questionnaires was to measure the business network influences on the attention for the environment, which formed the longitudinal backbone of the present study. They concentrated, however, on other aspects too. The survey of 2002 zoomed in on the relationship with government, evaluating, among others, the contact frequencies, perceived relevancy of exchanged information, and participation in public-private environmental agreements. Notably, the survey was supported by the Dutch Ministry of Housing, Spatial Planning, and the Environment (VROM). The goal of the overall project was to get more insight into the use of ICT to reduce the administrative consequences of environmental regulation (Bremmers *et al.*, 2003). For example, ICT can facilitate electronic governmental environmental reporting and, as such, reduce administrative burden. The survey of 2005 was used to get more insight into how firms encounter environmental issues, including an assessment of firm strategy and enabling organizational capabilities. It was measured to what extent internal environmental communication took place in order to enable the use of environmental knowledge and expertise available in the organization to develop environmental management capabilities.

5.2 Conceptual model

In line with the outside-in and inside-out perspective, Chapter 4 elaborated on the impact of business network actors as well as the importance of firm strategy and organizational capabilities that can enable the attention for the environment. Figure 5.1 shows the conceptual model.

It can be seen in Figure 5.1 that the stakeholder influences are split up into government, chain and network actors, and societal groups. Furthermore, a distinction is made between stakeholder pressures and environmental cooperation. Firms have to take care of environmental stakeholder expectations in order to obtain a *license-to-produce* from their stakeholders, such as government, buyers and suppliers, as well as the local community. The influence of different stakeholders on the firm will not be the same. Government is a dominant stakeholder for

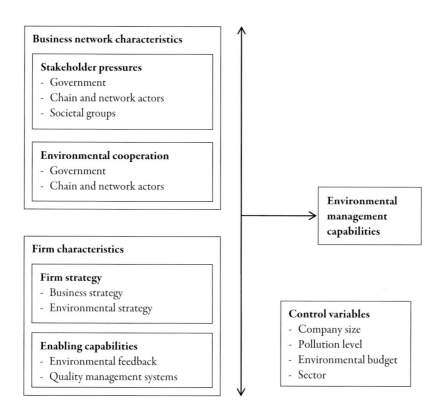

Figure 5.1. Conceptual model.

environmental management, because it can exert coercive pressure to provoke a reduction in environmental emissions in concordance with environmental regulation (Madsen and Ulhøi, 2001). It can also look for environmental cooperation in terms of public-private environmental voluntary agreements. Consideration of governmental requirements is essential to ensure conformity to legal requirements, which is, in turn, important to avoid business closure due to withdrawal of environmental permits. Environmental regulations can originate from different governmental levels, like the EU and the national government: it might be difficult for firms to comply with environmental demands from different governmental levels, especially if they are not sufficiently harmonized (Jordan and Liefferink, 2004). This was also indicated by several respondents during the semi-structured interviews (see Chapter 6). Furthermore, a distinction can be made between national and lower government, including various governmental bodies, such as provinces, municipalities, and water boards. A lack of harmonization of governmental policies and activities between these governmental levels and bodies might also hinder environmental compliance. For instance, environmental agreements are settled with the national government, while firms often have to deal with lower government for the execution in terms of environmental reporting duties and feedback.

Besides government, other stakeholders, affected by the firm's operations, will be interested in its environmental activities as well. For example, chain actors (e.g. buyers and suppliers) may want to ensure that the firm acts in concordance with environmental regulations to avoid illegal practices that could interrupt the supply-chain (Lippmann, 1999). They can also opt for environmental cooperation to collectively find opportunities to control and reduce the environmental impact. Examples are agreements on the supply of environmentally friendly produced raw materials or recyclable packaging. Network actors, like branch-organizations, can be important not only to increase the attention for the environment by diffusion of environmental information and knowledge (e.g. brochures and information bulletins), but also to facilitate public-private environmental cooperation through the accomplishment of environmental voluntary agreements with government. Societal groups, such as environmental organizations and local inhabitants, should be anticipated too. Taking care of their wishes is an important way of guaranteeing a so-called social *license-to-produce*, which is necessary for acceptance of the business activities causing pressure on the local environment, by people living near the firm (Kagan *et al.*, 2003).

Firm strategy reflects the long-term orientation of the firm on the establishment of a profitable balance between stakeholder wishes and the organizational resources. Figure 5.1 shows that it can be divided into business and environmental strategy. Environmental strategy is the part of the overall business strategy that considers the strategic position of the firm with respect to care for the environment. It can typically be classified as defensive versus proactive: the first implies a rather adverse attitude towards implementation of environmental measures, while the latter aims to proactively anticipate reduction of environmental impacts beyond legal requirements. An important driver of a proactive environmental strategy might be a business strategy that is directed at innovation in business processes and products. In this context, attention for the environment may add business value through implementation of advanced environmental processing techniques and meeting customer demands for environmentally friendly manufactured products. Hence, the development of environmental management capabilities might render positive effects, not only on the environment, but also on business performance in general.

Enabling capabilities are discerned, which are expected to enhance the development of environmental management capabilities. It is considered to what extent the knowledge and experience of people from different departments and organizational levels used for the environment. Furthermore, the development of environmental management capabilities might benefit from already implemented food quality and safety management systems. Not only expertise to deal with these systems could be an important source of knowledge, but also the use of existing managerial structures (e.g. regular meetings with the board for evaluation purposes) and operational activities (e.g. audits, training and education of personnel, and data gathering) could facilitate enrollment of environmental management capabilities (Karapetrovic and Willborn, 1998; Wilkinson and Dale, 1999).

The following control variables are discerned: company size, environmental pollution level, environmental budget, as well as food and beverage sector. Company size is important to consider, because large compared to small firms will, in general, have a higher level of pollution, which might feed the perceived urge to pay attention to the environment. In addition, they may attract more stakeholder interests than the small firms, which would increase the need to show environmental commitment (e.g. Bansal and Bogner, 2002). The environmental budget is also evaluated, because previous research indicates that small and medium-sized firms in particular perceive a shortage of financial resources as an important barrier to the implementation of environmental measures (Del Brío and Junquera, 2003; Hillary, 2004). However, large firms might equally have a lack of money to invest in the environment. Lastly, the Dutch food and beverage industry encompasses different sectors, like meat, dairy, and beverages. They can be characterized by the extent to which perishable (e.g. meat, fish, vegetables and fruit) and not or less perishable (e.g. animal concentrate and beverage) products are produced. The perishable products require quick handling (e.g. processing and distribution) to guarantee quality. This may have implications for the environmental pressure in terms of energy consumption for fast cooling (e.g. dairy products) and daily transportation to deliver fresh products to the supermarkets.

5.3 Operationalization

The research variables (or constructs) that are included in the conceptual model were operationalized for use in the Dutch food and beverage industry. The semi-structured interviews consisted of open questions in order to allow for discussion. A copy of the interview protocol can be found in Appendix A. The survey questionnaire used in 2002 and 2005 can be found in Appendix B and C, respectively. The measurement scales of the survey questions were carefully considered, because a good scale can enhance the validity and reliability of constructs (Fowler, 2002). Validity refers to the extent to which an answer to a question is a true reflection of the operational definitions (which are included in Table 5.1), whereas reliability is the extent to which respondents in comparable situations are likely to answer the questions in similar ways (Carmines and Zeller, 1979; Churchill, 1979; Fowler, 2002). This study applied mainly ordinal answer categories based on Likert scales (Churchill, 1999; Friedman and Amoo, 1999). The Likert scales used comprise 5 answer categories to enhance a balanced scale, while, at the same time, the meaning of the different alternatives is still feasible (Friedman and Amoo, 1999). Furthermore, nominal scales are used to collect factual data, such as the position of the respondent and the participation in various covenants. Ratio scales are employed for company size in terms of the number of employees. Lastly, open questions are also included for the respondents to provide additional information or comments.

Multiple-item constructs comprise of two or more questions to measure the research variable. The use of multiple-item constructs is preferred, because they are less sensitive to measurement errors (Churchill, 1979; Diamantopoulos, 1999). However, single-item constructs have also been used, if no valid and/or reliable multiple-item construct could be composed. The unweighted mean values of all constructs are calculated and included in the analyses, unless indicated otherwise. Table 5.1 shows the names and operational definitions of the research variables and questions used for measurement. The numbers in the table refer to the question numbers in the interview protocol (see Appendix A), the survey of 2002 (Appendix B) and 2005 (Appendix C). The remainder of this section discusses the constructed variables that are presented in Table 5.1.

The business network impacts are measured on a 5-point Likert scale with one end of the continuum denoting 'No influence at all' and the other 'A very strong influence'. Other empirical studies used comparable scales to measure stakeholder influences (Henriques and Sadorsky, 1996; Madsen and Ulhøi, 2001; Buysse and Verbeke, 2003; Sharma and Henriques, 2005). The relationship with the government is divided into the perceived overall impact of the government, frequency of contacts, quality of information provided, importance as information source, equality and dialogue of the relationship, as well as the contribution of governmental environmental policy to business goals. The frequency of contacts is measured as times that the firm has contact with lower governmental agencies, including the municipality, province, and water board, on environmental issues. An ordinal 5-point scale was used (1 = 'never', 2 = 'yearly', 3 = 'twice a year', 4 = 'four times a year', and 5 = 'monthly'). Furthermore, the respondents are asked to assess the other elements of the governmental relationship by indicating the extent to which they agreed with a number of statements, using a 5-point Likert scale ranging from 1 = 'I completely disagree' to 5 = 'I completely agree'. Lastly, public-private partnerships are assessed in terms of participation in environmental covenants.

The relationship with chain and network actors comprises the influence of other chain actors (i.e. buyers, suppliers, competitors, and consumers), the frequency of contacts with buyers and suppliers to arrange environmental agreements (measured on a scale similar to the frequency of contacts with lower governmental agencies), and environmental cooperation with buyers and/or suppliers (see Table 5.1). Respondents are asked to what extent the firm perceives help and support (e.g. exchange of experiences and knowledge) from another chain actor, such as a large buyer or supplier, for handling environmental issues. Furthermore, the importance of intermediaries in the network is measured, including the perceived environmental help and support from a branch-organization (e.g. provision of information on environmental regulation and/or environmental bulletins). Also the impact of banks and insurance firms on environmental management is considered. Last, the perceived pressure from societal groups, including environmental organizations and local inhabitants, is measured.

Business strategy is expressed in terms of what Miles and Snow (1978) call a prospector firm, i.e. one that wants to be first to introduce new products and acts in a market that can be characterized by fast-changing customer demands (see Table 5.1). Environmental strategy is reflected in top-management commitment to environmental management. It is also associated with the scope of environmental responsibilities attributed to the environmentally responsible manager in terms of its involvement in strategic and operational environmental decision-making. Business

Table 5.1. Operational definitions and indicators of the constructs and the measures used for the semi-structured interviews (the numbers refer to Appendix A) and the survey questionnaires in 2002 (Appendix B) and 2005 (Appendix C).

Constructs	Operational definitions and indicators	Interviews	2002	2005
Government				
Influence of government	Influence of the government on environmental management (in 2005 a distinction was made between EU, national, and local government)	D1-D2	9.4	7.5-7.7
Frequency of contacts with government	Frequency of contacts on environmental issues with the municipality, province, and the water boards		13.1-13.3	
Quality of information provided	Consistency of the information included in the environmental permits and a clear environmental policy towards the firm		11.1, 11.2	
Importance as information source	The importance to stay informed about legal requirements via a visit of the civil servant to the firm and vice versa, as well as a fixed contact person at lower governmental agencies		12.1-12.3	
Equality and dialogue of the relationship	The involvement in environmental decision-making at lower government and if there is an open dialogue and informal communication on environmental policy goals		11.3-11.5	
Contribution to business goals	The extent to which governmental environmental policy goals contributes to business goals		11.6	10.5
Public-private partnerships	Sum of participation in the Packaging Covenant, Long-Term Agreement Energy, Energy Benchmarking Covenant, Integral Environmental Task		7.1-7.4	8.1-8.4
Chain and network actors				
Influence of buyers and suppliers	Influence of buyers and suppliers on environmental management		9.1, 9.2	7.1, 7.2
Frequency of contacts with buyers/suppliers	Frequency of contacts to arrange environmental agreements with buyers and/or suppliers between the following organizational levels/departments: CEO, purchase, sales, marketing, R&D, production, logistics, quality		8.1-8.8	
Cooperation with buyers and suppliers	Environmental cooperation with buyers and suppliers	B4, D4	10.1, 10.2	10.3, 10.4
Influence of competitors	Influence of competitors on environmental management		11.3	7.3

Table 5.1. Continued.

Constructs	Operational definitions and indicators	Interviews	2002	2005
Branch-organizations	Branch-organizations that provide help for environmental issues (in 2005, their influence on environmental management was requested)		5	7.4
Chain captain	Chain captain that provides help for environmental issues		6	
Influence of banks/insurance firms	Influence of banks/insurance firms on environmental management	D3		7.11
Influence of consumers	Influence of consumers on environmental management			7.9
Societal groups				
Influence of local inhabitants	Influence of local inhabitants on environmental management	D5	9.6	7.10
Influence of environmental organizations	Influence of environmental organizations on environmental management	D5	9.5	7.8
Business strategy				
Prospector firm	The firm wants to be first to introduce new products and customers constantly ask for new products	D6		2.1, 2.2
Environmental strategy				
Top-management commitment	Board of directors is actively involved in environmental management issues	E3		6.8
Influence of environmental coordinator	The environmental coordinator has a strong impact on strategic and organizational environmental issues	C1-C2		6.4, 6.5
Environmental feedback				
Environmental communication	Organizational culture that stimulates sharing of environmental ideas, environmental issues can be adequately communicated to higher management, environmental information is shared among employees	D3		6.1-6.3
Involvement of different departments	Different departments are involved in environmental decision-making and environmental problems are collectively dealt with			6.6, 6.7
Influence of employees	Influence of employees on environmental management			7.12
Quality management systems				
Integration environ. issues in quality management	Integration of environmental issues in food quality management systems	C5		10.1

Variable	Description	Code		
Integration environ. issues in HRM	Integration of environmental issues in human resource management (HRM) systems	C5		10.2
Environmental management capabilities				
Strategic, operational, chain and cradle-to-cradle capabilities	Sum of the availability/implementation of 2 strategic (an environmental strategy and action program), 8 operational (environmental audit, environmental training of employees, environmental database, regular measurement of environmental impact, internal environmental information processing, environmental information collection for internal environmental care, to check environmental emissions, and to evaluate environmental measures, and 3 chain and cradle-to-cradle (information collection for chain-oriented environmental care, environmentally friendly product (re)design, and to exchange with buyers/suppliers) capabilities. After rescaling (*5/13) expressed on a scale from 0 (= lowest performance) to 5 (= highest performance)	B4, EI	3.1-3.6, 4.1-4.7	4.1-4.6, 5.1-5.7
Benefits				
Image	The extent to which environmental management contributed an environmentally friendly image	E2		11.1
Marketing	The extent to which environmental marketing opportunities were created	E2		11.2
Control variables				
Firm size	Number of employees (source: Dutch Chamber of Commerce)	B3	Chamber of Commerce	Chamber of Commerce
Pollution level	Mean value of the most important perceived pollution issues: soil pollution, noise, water pollution, smell, air pollution, hazardous substances, waste, and energy consumption (the latter was asked in 2005 only)	B5	2.1-2.7	3.1-3.8
Environmental budget	Availability of a sufficient environmental budget			10.3
Sector	Classification of firms into sectors that produce perishable products (i.e. bakery, vegetables and fruit, dairy, and meat) versus less or non-perishable products (i.e. animal feed, beverage, grain mill products, and other sectors, such as coffee and tea)	Chamber of Commerce	Chamber of Commerce	Chamber of Commerce

Table 5.1. Continued.

Constructs	Operational definitions and indicators	Interviews 2002	2005
General information			
Position of the respondent	Main position of the respondent (environmental coordinator, quality manager, board member, etc.)	A1-A3	—
History firm	Date of establishment and core business	B1-B2	
ISO14001 certification	Formal certification of ISO14001 (source: Dutch official certification body SCCM)	B6	SCCM

strategy and environmental strategy are measured based on statements that are answered on 5-point Likert scales (1 = 'I completely disagree' to 5 = 'I completely agree').

The measured enabling capabilities comprise environmental communication, the involvement of people from different departments in environmental decision-making, as well as the impact of employees on environmental management. Environmental communication is measured by statements on sharing environmental issues among employees and with higher-level management. The involvement of different departments reflects the attention for environmental issues in the organization.

In order to determine the effect of integrated quality management systems, it is asked to what extent environmental and food quality and safety management were integrated. Also the level of integration of environmental and human resource management systems is measured.

The adoption of environmental management capabilities is based on the implementation of various environmental management activities. A distinction is made between reasons for environmental information gathering (e.g. to check environmental emission or to evaluate environmental measures) and the implementation of an environmental management system. The research variable is operationalized as the sum of 13 binary variables (yes/no) listed in Table 5.1. In doing so, an operationalization strategy is chosen between empirical studies that used a single-item construct to measure attention for environmental management (e.g. Henriques and Sadorsky, 1996; Roy *et al.*, 2001) versus those that used over 50 items (e.g. Sharma and Vredenburg, 1998). The included variables are based on a set of environmental management principles that are developed by Dutch employer organizations (VNO/NCW) together with the Dutch government (Braakhuis *et al.*, 1995; RMK, 1997). Notably, these principles are in line with ISO14001 guidelines on environmental management system implementation (Martin, 1998; Netherwood, 2004). After summation, the environmental management performance is rescaled by multiplying with a factor 5/13, resulting in a scale from 0 (= lowest performance) to 5 (= highest performance). For the analysis a distinction is made between strategic (i.e. environmental strategy and action program), operational (e.g. information collection for checking emissions and keeping an environmental database), and chain and cradle-to-cradle capabilities as well. The latter involves activities that enhance attention for the environment from a life-cycle approach, such as environmentally friendly product (re)design and information collection for environmental care at the chain-level.

For explorative reasons, it is also asked to what extent environmental management contributes to an environmentally responsible image and marketing opportunities. The use of perception scales is in line with previous studies that employed similar measures (Halkos and Evangelinos, 2002; Sroufe, 2003).

The control variables consist of company size, pollution level, environmental budget, and the sector. Firm size is expressed as the number of employees. Multiple other studies on environmental management used this proxy for size (Fryxell and Szetto, 2002; Halkos and Evangelinos, 2002; Branzei *et al.*, 2004), although turnover is also commonly used (Buysse and Verbeke, 2003; Melnyk *et al.*, 2003). The present study obtained data on the number of employees from a reliable secondary data source, namely the Dutch Chamber of Commerce, which could not be done for turnover.

The pollution level is measured as the perceived significance of the environmental impact caused by the firm. Because it is hard to obtain physical data on environmental emissions, in particular at the firm level, other studies used the perception of respondents as a proxy for the level of environmental pressure, as well. Christman (2000) asked, for instance, to indicate the one most important environmental issue that had an effect on business. Respondents are asked to assess the significance of several pollution issues for their plant-site, such as the importance of taking care of noise production, soil pollution, and release of hazardous substances. A 5-point Likert-scale is used (1 = 'Not important at all' to 5 = 'Very important'). The level of pollution is calculated as the mean value of the most important issues. Madsen *et al.* (1997) also employed a perceptual measure, asking respondents to assess the level of environmental emissions. They found that respondents indicated lower levels of emissions than could be expected from pollution figures at industry level. The indicated significance is therefore carefully checked with the important pollution issues found for the Dutch food and beverage industry as a whole (see Section 7.3.1, 7.4.1, and 7.5.1).

The environmental budget is considered by asking the respondents to what extent they have sufficient financial resources to invest in the environment. It is measured on a 5-point Likert scale (1 = 'I completely disagree' to 5 = 'I completely agree').

Differences in food and beverage sectors that may affect attention for environmental management is considered in terms of production of perishable and less or non-perishable products. Bakery, vegetables and fruit, dairy, and meat are classified as perishable food sectors, although they deliver also processed products, such as canned meat and fruit). The animal feed, beverage, grain mill products, and other sectors (e.g. coffee, tea, and cocoa) are classified as less perishable.

General background information is gathered. The position of the respondent is asked (e.g. CEO, environmental coordinator and/or quality manager), while only the semi-structured interviews pay attention to the history of the firm as well. ISO14001 certification is considered. Information on this is obtained from the central public-private body responsible for registration of environmentally certified organizations in the Netherlands (SCCM, see: www.sccm.nl). Certification was checked one year after the survey in 2005, in November 2006. In doing so, preparation for certification (while not being certified yet), which may contribute positively to attention for environmental management, is also taken into account.

The unit of analysis is the plant or business unit (in case of small firms). This choice is underpinned by the fact that the urge to take care of the environment will be affected by local circumstances, such as the distance to a local community or nature conservation resources (Klassen and Whybark, 1999; Sharma and Henriques, 2005). Noise and smell problems may, for instance, be important to deal with because of complaints from the local community. Consideration of their wishes is necessary to guarantee a social license-to-produce. In reaction to negative effects on nature conservation, the government may, for instance, impose increased environmental requirements.

5.4 Propositions

5.4.1 Business network

The netchain concept discusses the connection of a firm with other network actors that can be found along the horizontal and vertical axes of the business network (see Section 4.4.2). The horizontal axis refers to the interaction with competitors, government and branch-organizations, while the vertical axis represents the connections with other chain actors, such as buyers, suppliers, and consumers. Government wants firms to take care of their environmental impacts. As mentioned in Section 5.2, two extremes of governmental environmental policy are command-and-control versus public-private environmental partnerships (Prakash and Kollman, 2004). In case of command-and-control, government prescribes environmental emission reduction targets and environmental measures to clean up (e.g. emission reduction filters), whereas public-private partnerships refer to voluntary environmental agreements between the government and firms to reduce their environmental pressures. Of course, other policy modes can be found in between these two extremes, like market-based initiatives (e.g. tradable emission permits) and mandatory environmental information disclosures, like governmental environmental reporting (Sinclair, 1997; Prakash and Kollman, 2004). The extent to which the government can enforce firms that demonstrate an adverse attitude towards environmental regulations will depend on government's ability to exert coercive pressure, threatening with financial penalties and withdrawal of environmental permits to avoid illegal practices. Imposing environmental requirements might be necessary to overcome resistance to taking care of environmental impacts particularly if measures are perceived as costly only (Van Snellenberg and Van de Peppel, 2002). The mutual consideration of public and private environmental interests during covenanting processes is expected, however, to increase the commitment of firms to meet the agreed environmental targets on a voluntary basis (Glasbergen, 1999). While the voluntary character is sometimes criticized, the typical involvement of branch-organizations is expected to contribute not only to the perceived legitimacy and thus acceptance of the covenant, but also increased pressure from the business network to take part in the covenant. As such, environmental agreements may positively influence the environmental attitudes adhered in the business network. The participation in an environmental covenant is therefore not only expected to increase attention for the environmental issues addressed in the agreement, but also to stimulate the development of environmental management capabilities in general.

P1 The participation in environmental voluntary agreements will be positively related to the adoption of environmental management capabilities.

The legal power of government to enforce firms to clean up is associated with Granovetter's (1992) structural rather than relational embeddedness, since it reflects the position of government as a network actor standing above all firms creating a regulatory level playing field. Relational embeddedness refers to the strength of business network contacts, which can be measured in terms of perceived quality of the governmental information and dialogue on environmental issues (see also Table 5.1). In this perspective, it is expected that direct contacts with lower government are important to obtain valuable environmental information on environmental regulations and the consequences these regulations for the firm. In particular in the small firms, this information may help overcome limited own environmental resources to take account of this (Hillary, 1999; Del Brío and Junquera, 2003; Hillary, 2004). However, also medium-sized and large firms might benefit from direct contacts with lower governmental agencies for this reason. In sum, it is expected that the contact intensity with lower government will be positively related to the development of environmental management capabilities.

P2 The contact intensity with lower government will be positively related to the adoption of environmental management capabilities.

At the horizontal level of the business network, environmentally active branch-organizations are expected to be important to stimulate firms to 'green' their business. They act as an intermediary between government and the firms. They can be involved in the accomplishment of environmental covenants and try to influence proposed or future environmental regulations to the benefit of their sector (or branch). In addition, branch-organizations can play an important role in establishing environmental agreements at the sector level, such as on environmental quality standards. The betweenness-centrality concept, as it is discussed by Rowley (1997), states that intermediaries have the exclusive opportunity to influence the information that is exchanged between the actors to which they are linked (see Section 4.2.1). Taking this into account, branch-organizations are expected to have the power to increase environmental attention in firms by stressing the urge to conform to legal environmental demands and adhering to ambitious environmental covenants. It is proposed that, if branch-organizations pay attention to the environment, it will diffuse environmental awareness at the horizontal level of the business network and, consequently, stimulate the development of environmental management capabilities.

P3 Environmental support from an intermediary, such as a branch-organization, will be positively related to the adoption of environmental management capabilities.

At the vertical axis of the business network, perceived impacts from other chain actors, such as buyers and suppliers, might enhance attention for the environment. Their interest in controlling environmental pressures can be associated with safeguarding the proper functioning of the chain as a whole, including the regular supply of raw materials, which might be hindered in case of closure of a firm due to an environmental scandal (Lippmann, 1999). In this context, large buyers in particular may induce environmental standards, carrying out site audits to check compliance. If the firm does not act in concordance with the requirements, it will be replaced by another supplier, especially when there is not a strong resource interdependency (Sharma and Henriques, 2005). In fact, Holt (2004) argues that firms increasingly impose environmental selection criteria on their suppliers, in order to satisfy the environmental wishes of their own customers. It is therefore expected that increased attention for the environment among chain actors, and in particular downstream in the chain (e.g. buyers and consumers), contributes significantly to the urge to develop environmental management capabilities.

P4 Pressure from chain actors to pay attention to the environment will be positively related to the adoption of environmental management capabilities.

5.4.2 Firm strategy

Business strategy will have implications for the attention paid to the environment. Miles and Snow (1978) identify prospectors as firms with the most innovative business strategy compared to defenders and analyzers[11]. Prospectors look for new products and marketing opportunities and they try to innovate by regularly redefining the focus of their business on specific goods and services, as well as target markets and segments. They use flexible production technologies that enable them to adapt quickly to new product demands. By contrast, defenders concentrate on a limited set of products, which they try to manufacture at the lowest cost possible. Analyzers can be interpreted as a unique combination of prospectors and defenders: they aim to minimize risk by low-cost production of core products, while maximizing opportunity by looking for new market chances. It is expected that prospectors compared to both defenders and analyzes will sooner implement new clean technologies and respond to customer wishes for environmentally responsible product (re)design. In this context, Aragón-Correa (1998) emphasizes their willingness to enhance technological leadership and capacity to quickly adapt to new and improved technologies. Furthermore, Christmann (2000) found that prospectors benefit from paying attention to the environment, because it is in line with their business strategy to be first to launch environmentally friendly products and carry out environmental technologies. It is expected that prospectors will be especially interested in cradle-to-cradle environmental management capabilities, such as environmentally friendly product (re)design on the use of recycled and recyclable products and components.

---

[11] In addition, they identify reactors, which represents a rather unstable strategy, because it lacks a consistent organizational adaptation to market demands. It is sometimes referred to as a 'residual strategy' (see also: Miles *et al.*, 1978).

*P5 A prospector strategy will be positively related to the adoption of environmental management
capabilities, in particular cradle-to-cradle capabilities (e.g. environmentally friendly product
(re)design).*

Environmental issues are often dealt with by middle management, such as the head of
the production department or plant managers (depending on the size of the firm). They
should make sure that the firm acts in legal compliance. Fryxall and Vryza (1999) found
that environmental responsibilities are often dealt with in close cooperation with the legal
department in large firms. However, the positive effects of attention for the environment on
business performance, such as less cost for energy usage and waste disposal, might gradually
attract increased top-management interest. What started as incremental environmental
changes, driven by a compliance-oriented environmental strategy, might typically evolve into
increased strategic attention to take care of the environment (Cramer, 1998; Blomquist and
Sandström, 2004). If top management is convinced of the environmental benefits for the firm,
environmental issues are likely to get strategic managerial priority to stress environmental
excellence. The strategic commitment is, in turn, expected to further enhance attention
for the environment through the allocation of additional resources (e.g. financial budget).
Accordingly, the following proposition is formulated.

*P6 Strong top-management commitment to the environment will be positively related to the
adoption of environmental management capabilities.*

Top-management commitment is rhetoric, if their environmental promises are not followed
by real attention for the environment (Rhee and Lee, 2003; Peattie and Crane, 2005). It
is argued that their environmental commitment should be accompanied by clearly visible
environmental measures, including the appointment of an environmental manager to stimulate
organizational changes in line with environmental strategy (Prakash, 2001; Zutshi and Sohal,
2004a). This environmental manager can be employed to increase environmental awareness
in general and to ensure that people are informed about the organizational consequences
of new environmental strategies. Interestingly, recent empirical investigations underpin the
importance of an environmental manager to enhance the implementation of environmental
measures (Blomquist and Sandström, 2004; Zutshi and Sohal, 2004a). This can be explained
by the fact that an environmental manager supports the translation of environmental ambitions
in practice and can identify environmental problems at an early stage to take (new) strategic
and/or operational measures. The following proposition is therefore defined.

*P7 A strongly perceived impact of the environmental manager on operational and especially
strategic environmental issues will be positively related to the adoption of environmental
management capabilities.*

5.4.3 Enabling capabilities

Employees are important for identifying environmental problems and taking immediate care in case of a chemical incident or heavy contamination of surface water (Zobel and Burman, 2004). Their knowledge and experience is essential to identify critical success factors for the implementation of environmental measures, such as the extent to which new clean technologies can be embedded in existing organizational routines. Consideration of their environmental expertise, captured in knowledge and experience, is important to ensure changes in the organization to the benefit of the environment (Boiral, 2002; Subramaniam and Youndt, 2005). It is crucial for the development of environmental management capabilities, since it will provide for valuable feedback, learning effects, and, in turn, result in improved environmental practices. In this perspective, it is suggested that well-established communication channels to enable environmental feedback in the organization will enhance the adoption of environmental management capabilities.

P8 Higher levels of horizontal (e.g. at the same organization level) and vertical (i.e. bottom-up and top-down) communication will be positively related to the adoption of environmental management capabilities.

Food and beverage firms are legally obliged to ensure food safety according to different (inter)national standards (see Section 3.3.3). Consequently, many firms have implemented care systems to ensure food quality and safety. The capabilities related to the implementation of these management systems are expected to show similarities with environmental management capabilities, which can be used to formulate strategic environmental targets, implement environmental measures, and review business performance (Karapetrovic and Willborn, 1998). It is suggested that existing managerial structures, such as regular reporting to top-management and auditing of the production process, can be extended to include environmental issues as well. However, in order to overcome limited managerial flexibility in the individual management areas, overly tight integration might be disadvantageous (Karapetrovic and Willborn, 1998; Wilkinson and Dale, 1999). For example, the required frequency of audits may be higher for food quality and safety management compared to environmental management. In summary, it is proposed that the availability of quality care systems can enhance the development of environmental management capabilities.

P9 The availability of (other) care systems (e.g. on food quality and safety) will be positively related to the adoption of environmental management capabilities.

5.5 Data gathering

Data are gathered through semi-structured interviews and survey questionnaires. The interviews are carried out before and after the survey in 2005. The interviewees are selected from the questionnaire surveys, in which cooperation was requested in an additional interview.

In total, fifteen interviews were scheduled. However, one respondent appeared to have left the firm: his successor turned out to be unable to answer the questions adequately. Furthermore, one meeting was cancelled, because the interviewee was ill. Hence, a total of 13 firms have been visited. Similar procedures are used for all interviews. They are prepared based on an interview protocol to keep the discussions with respondents focused on relevant topics, which, in turn, contributes to the reliability of the data obtained (Eisenhardt, 1989; Yin, 2003). The meetings lasted 60-90 minutes. A semi-structured interview protocol is used (see Appendix A). This means that no strict order of topics is followed, since this was left to the preference of the interviewees. All interviews are recorded and translated on paper. Additional information is, if necessary, gathered from auxiliary sources (e.g. environmental reports) or requested by e-mail. The respondents received a feedback report within four weeks to check if their answers were quoted correctly.

In 2002, seven experts from government and branch-organizations are consulted to assess the content validity of the survey questionnaire. In 2005, the content validity is checked during the interviews. Questions are revised, if necessary. General guidelines, such as a clear and easy to read lay-out and short instructions, are taken care of to improve the quality of the questionnaires (Fowler, 2002). Both survey questionnaires focus on the business network impacts, while special attention is paid to governmental relationships in 2002 and to firm strategy and enabling capabilities in 2005. In order to examine differences between micro (5-10 employees), small (10-50), medium-sized (50-250), and large firms (>250 employees), the survey of 2002 is sent to Dutch food and beverage firms with five or more employees[12]. To get a more homogeneous sample with respect to firm strategy and enabling capabilities, the survey of 2005 is sent to medium-sized and large Dutch food and beverage firms only. The demarcation of the study population is based on the classification codes of the Dutch Chamber of Commerce (so-called 'BIK codes').

Table 5.2 includes all sectors of the Dutch food and beverage industry. The addresses of the firms are obtained from the Dutch Chamber of Commerce. The mailings consisted of a cover letter, the questionnaire, a pre-paid return envelope, and a short list of definitions (the latter is presented in Appendix D). The package is addressed to the manager responsible for environmental issues (e.g. environmental coordinator, board member, and/or quality manager). The cover letter contained background information on the research. Anonymity in participation is ensured in order to enhance the response rate. In 2002, no follow up is sent after the first mailing, because the response was already satisfactory. As stated previously, the survey is carried out on behalf of the Ministry of VROM, which might have enhanced cooperation in the survey. Moreover, the scope of the questionnaire included evaluation of administrative loads of governmental regulation, which might also have encouraged firms to participate. A slightly different procedure is followed in 2005: since the Ministry was not

---

[12] This classification is in line with EU definitions on micro, small, medium-sized, and large firms regarding the number of employees (see: ec.europa.eu/enterprise/).

Table 5.2. Included firm categories based on the classification codes for the food and beverage industry used by the Dutch Chamber of Commerce.

BIK-code		BIK code	
151	Slaughterhouses and meat processors	156	Grain mill and starch processors
152	Fish processors	157	Animal feed producers
153	Vegetables and fruit processors	158	Bakery factories and others
154	Producers of fat and oil	159	Beverage producers
155	Producers of dairy products		

involved, it was expected to be harder to yield a satisfactory response. In 2005, firms could choose to get an individual feedback report (see Appendix E). Following recommendations by James (1994), a small monetary incentive is provided as well, in order to increase the willingness for cooperation. Two sequential mailings are carried out. After the first mailing, the reasons for non-response are investigated randomly by telephone at approximately 50% of the non-responding firms. It is used to identify the name of the person to include in the follow-up mailing, in order to increase the response (Dillman *et al.*, 1974). The second mailing is carried out four weeks later.

5.6 Statistical analyses

The statistical package SPSS version 12.0.1 is used for the analyses. The statistical analyses include both parametric (e.g. linear regression analyses) and non-parametric techniques (e.g. Spearman rank correlations). This is necessary to deal with the fact that the majority of the research variables is measured on an ordinal scale (e.g. using Likert categories). The remainder of this section discusses the statistical techniques carried out.

5.6.1 Response analysis

The representativeness of the samples is tested with respect to sector and company size (i.e. number of employees), using a Chi-square (χ^2) test. In contrast to 2002, the data of 2005 are collected in two batches, which allows for an additional non-response analysis based on a comparison of the firms that replied after the first and second mailing. In doing so, it is assumed that the late respondents (i.e. that replied after the second mailing) have similar characteristics to non-respondents (Armstrong and Overton, 1977). An Anova t-test is used to examine differences in company size, while all other research variables are compared, using a Kruskal-Wallis test for non-parametric data analysis.

5.6.2 Validity and reliability of the constructs

A distinction can be made between reflective and formative constructs. The first comprise variables that are dependent on the meaning of the underlying dimensions of the construct, while the latter consist of variables that are defining the meaning of the constructs itself (Diamantopoulos, 1999; Diamantopoulos and Winklhofer, 2001; Jarvis *et al.*, 2003). As a rule of thumb, the items of a reflective construct are inter-changeable, whereas the elimination of one or more items from a formative construct will alter its meaning. The following formative constructs are used in the present study: frequency of contacts with buyers and suppliers, frequency of contacts with lower governmental agencies, environmental cooperation with buyers/suppliers, environmental pollution level, and the adoption of environmental management capabilities.

A distinction can be made between content, convergent, discriminate, and nominal validity. The first can be assessed for both types of constructs, while the latter three can only be applied on reflective constructs. Content validity (or: face validity) is the degree to which the construct is focused on a specific domain (Carmines and Zeller, 1979). The present study has taken care of this by means of a literature review (see Chapter 2 and 4), an adequate conceptual design, and the evaluation of the questionnaires by experts (Churchill, 1979; 1999; Fowler, 2002). Convergent validity refers to the correlation between the variables attributed to one construct. It tests whether they belong to the same concept. Factorial data reduction (principle component with varimax rotation, eigenvalue ≥ 1) is used to assess convergent validity. Discriminant validity is the degree to which the constructs are distinctively different measures compared to each other. This is checked by means of Spearman correlation coefficients between the constructs, which should be lower than 0.80 (Hair *et al.*, 1998). Nominal validity includes a check on the significant relationship of the construct with the dependent variable(s). The results of the regression analyses are used for this.

The reliability of the reflective constructs is assessed with Cronbach α, which measures the internal consistency of the items, thought to reflect a single construct. As a lower level of acceptability it is common to use 0.70, although 0.60 is acceptable in explorative research (Hair *et al.*, 1998). The present study evaluates the reliability of the formative constructs, since a concept version of the survey questionnaire was completed during the semi-structured interviews.

5.6.3 Baseline statistics and correlations

In line with the EU definitions on micro, small, medium, and large firms, a distinction is made between the following size categories: 5-10, 10-50, 50-250, and more than 250 employees (see also footnote at Section 5.5). The availability of environmental management capabilities is evaluated for these groups. Significant differences in percentages are manually tested, following the statistical procedure outlined by Churchill (1999). MS Excel is used to carry out the additional analysis. As part of the baseline statistics, the mean values and standard deviations are calculated for the research variables. The Kruskal-Wallis test is used to test differences between multiple groups (i.e. size-categories), while the Mann-Whitney test is applied to test between two groups. Both tests are suitable for non-parametric data analysis. Significant differences between size-categories in the implementation of different elements of environmental management capabilities are tested by comparing sample proportions (Churchill, 1999). Furthermore, Spearman rank correlation coefficients for non-parametric data are evaluated. Missing values are deleted listwise.

5.6.4 Regression analyses

To find significant determinants for the adoption of environmental management capabilities all research variables are entered as independent variables in the regression models (see also Table 5.1). The analysis is split up for micro and small versus medium-sized and large firms: missing values are deleted listwise. The regression models were validated based on a check of the predictive power, the existence of multi-collinearity, and the model residuals (Onwuegbuzie and Daniel, 2003). The predictive power of the linear regression models is measured by the coefficient of determination (R^2), which is the proportion of total variance explained by the model. Checks for multi-collinearity are based on Spearman correlations (coefficients should be lower than 0.80) and the variance inflation factor (VIF). The commonly applied maximum threshold value for VIF is 10 (Hair *et al.*, 1998). The distribution of the standardized residuals is checked with the Kolmogorov-Smirnov (K-S) test. Because it checks deviation from a normal distribution, non-significant outcomes indicate a normal distribution (Field, 2003).

Referring forward to the results, it can already be mentioned that the micro and small firms turned out to have implemented only few elements of the environmental management capabilities (see Section 7.3.1). A binary logistic regression is therefore carried out in this case. The sample is divided into firms that implement none versus one or more items of the environmental management capabilities. As recommended by Field (2003), inspection of multi-collinearity is based on procedures similar to those used in linear regression. Two differences in the interpretation of the model results should be clarified. First, it presents parameter estimates that indicate the association between the adoption of environmental management capabilities and the other research variables in terms of likeliness or odds-ratios. Second, the Nagelkerke pseudo R^2 evaluates the predictive power: it can be interpreted in a similar way to R^2 in linear regression analysis.

5.6.5 Cluster analysis

A cluster analysis is performed to get more insight into the joint impact of the business network, firm strategy, and enabling capabilities. Based on recommendations by Ketchen and Shook (1996), the significant predictors of the regression analyses are used as clustering variables. The control variables are, however, not included, since they are used to validate the clusters. A hierarchal cluster analysis (Ward's cluster method) is applied to identify the seed points of a succeeding non-hierarchical cluster analysis (K-means). This two-stage procedure contributes to robust cluster groups (Milligan, 1980; Ketchen and Shook, 1996). Ward's clustering method is applied, because it enhances equal cluster groups with respect to the number of observations. The number of adequate clusters is visually checked with a dendogram. Lastly, the cluster groups are validated based on significant differences in all research variables (including the control variables).

5.6.6 Longitudinal analysis

A longitudinal analysis is carried out to get insight into significant changes in business network impacts and attention for environmental management capabilities in medium-sized and large firms. The mean values and Spearman correlations are evaluated as well. A regression analysis is performed, including all medium-sized and large firms. A dummy for the year of measurement is included to anticipate differences between 2002 and 2005.

5.7 Concluding remarks

This chapter has presented the conceptual model and propositions. It elaborated on the performed semi-structure interviews and survey questionnaires used in 2002 and 2005. It discussed the operationalization of the conceptual model into research variables (or constructs) that are used to test the propositions in the Dutch food and beverage industry. The backbone of the two surveys is the measurement of business network impacts. In 2002, micro (5-10 employees), small (10-50), medium-sized (50-250), and large (>250 employees) firms are included. Special attention is paid to the governmental relationship in terms of perceived quality of provided environmental information by and the intensity of contacts with lower government. The survey of 2005 is conducted among medium-sized and large food and beverage firms, because it pays special attention to firm strategy and enabling capabilities. Finally, this chapter elaborated on the univariate and multivariate statistical techniques which are used to analyze the data.

6. Interview results

This chapter evaluates the results of the semi-structured interviews that are carried out during on-site visits. Section 6.1 presents a baseline description of the firms that participated. Section 6.2 evaluates the business network impact on the attention for environmental management capabilities. In line with the research model, a distinction is made between the government, chain and network actors, and societal groups. Section 6.3 discusses the organizational capabilities related to environmental management. It concentrates on the strategic motives to pay attention to the environment, the organizational structure and environmental responsibilities, as well as the embeddedness of environmental care in the organization. Furthermore, it discusses the integration of different care systems and internal environmental information processing, including environmental indicators and investment costs. Section 6.4 provides the concluding remarks of this chapter and it links the empirical finings to the propositions.

6.1 Baseline description of the firms

In total, thirteen semi-structured interviews are carried out. Table 6.1 shows an overview of the main characteristics of the medium-sized (50-250 employees) and large (>250 employees) food and beverage firms that are interviewed. They are active in different sectors, including beverages, meat, bakery, vegetables and fruits, and animal feed. Firm 6 belongs to the category other, as well as firm 12, which produces food and beverage products. The large firms often have multiple plants (see Table 6.1).

In line with the expectations, energy consumption, handling of waste water, as well as organic waste are mentioned as the most important environmental issues. The attributed importance depends not only on the production processes (e.g. baking bread versus meat processing), but also on the firm's location. The interviewee of firm 2 indicates that smell from their waste water disposal is (still) an issue, despite the fact that advanced reduction technologies reduced it by 98%. The urgency is due to complaints from the local community: *At our former production location, smell from waste water disposal was not an important issue. At our new plant-site, however, people keep complaining.* In line with this, the respondent of firm 5 states: *Smell is not an urgent issue, because we are located at a large industrial site, but I do know other bakery factories, which had to move due to complains from the local community.* Furthermore, the importance of environmental issues is associated with economic motives, such as reduction of energy consumption or organic waste to save money, like in the meat processing firms (firm 7, 8, and 9).

Three firms are ISO14001-certified, although in case of firm 8 and 12 it includes only parts of the organization (see Table 6.1). Firm 2 is certified, because *it guarantees that we keep paying attention to the environment and it is also a positive sign to the government, since it shows our commitment.* Despite the fact that many respondents stated working in concordance with ISO14001 (firm 1, 4, 10, and 12), they indicate that there is a lack of economic incentives to get official certification.

Table 6.1. Characteristics of the firms and the position of the respondents[1].

Firm	Sector	Main product	Size	Number of production locations	Main environmental issues
1	Beverage	Beer	Medium	Single	Energy, waste water
2	Beverage	Beer	Large	Single	Energy, smell (from waste water disposal)
3	Dairy	Various cheese products	Medium	Multiple	Energy, waste water, noise, smell
4	Dairy	Various dairy products	Large	Multiple	Energy, waste water
5	Bakery	Bread and pastry	Medium	Single	Waste disposal
6	Other	Bakery ingredients	Large	Single	Waste water, packaging
7	Meat	Pig meat	Medium	Single	Energy, organic waste, noise, smell
8	Meat	Pig and beef meat	Large	Multiple	Energy, organic waste, waste water
9	Meat	Various meat snacks	Medium	Single	Energy, packaging, waste water, smell
10	Grain mill and starch	Various potato products and starch	Large	Multiple	Waste water, noise, chemical substances
11	Vegetables and fruit	Various potato products	Large	Multiple	Energy, waste water, organic waste, smell
12	Other	Sugar and potato products, alcohol	Large	Multiple	Energy, organic waste, smell
13	Animal feed	Animal feed	Medium	Single	Noise, dust, energy

[1] All interviews are carried out in 2005, except for firm 6, 9, 11, and 13, which are visited in 2007.
[2] Half of the number of plants are ISO14001-certified.
[3] Only parts of the organization.

Retail chains are generally mentioned as the most important buyers (see Table 6.1). The breweries (firm 1 and 2) also sell to cafés, hotels, and restaurants. Notably, the majority of the firms concentrates next to the Netherlands in international markets, especially Belgium and

ISO14001-certified	Main buyers	Purchase market	Position of respondents
No	Retail and horeca (e.g. cafés, hotels, restaurants)	Mainly Dutch	Board member
Yes	Retail and horeca	(Inter)national	Manager HRM and environment
No	Retail and horeca	(Inter)national	Environmental coordinator
No	Retail	(Inter)national	Corporate environmental officer
No	Retail	Dutch	Board member/owner
No	Bakery factories	(Inter)national	Manager raw materials
No	Other food processors	(Inter)national	Manager plant-site
Yes[2]	Retail, food processing industry, horeca	(Inter)national	Corporate environmental officer
No	Retail	(Inter)national	Manager TQM
No	Other food processors	(Inter)national	Manager TQM
Yes	Fast food industry	(Inter)national	Corporate manager environmental affairs
Yes[3]	Other food processors	(Inter)national	Corporate manager environment and quality
No	Farmers and small animal feed firms	Mainly Dutch	Manager TQM

Germany, but also Canada and the U.S. All respondents are involved in environmental issues as part of their formal responsibilities. In some firms, the interviewees are member of the board, as in the case of a medium-sized brewery (firm 1) and a bakery factory (firm 5).

6.2 Business network impact

6.2.1 Government

All firms have contacts with provinces and/or municipalities, as well as water boards for environmental permits. The role of provinces and municipalities is, however, stressed: nine firms deal mainly with the municipality and four with the province. Several respondents, such as in firm 4 and 10, have regular contacts with both for different environmental issues. Furthermore, it appears that firm 9 and 13 deal with regional governmental agencies, which act on behalf of multiple municipalities in one region. The remainder of this section discusses the respondents' comments made on the relationship with lower government.

Support and quality of the relationship

It is stressed that the quality of the relationship with the lower government, as perceived by the respondents, strongly depends on the governmental civil servants involved: not all are perceived as equally knowledgeable and/or prepared to discuss environmental issues (firm 2, 4, 7, 9, and 10). The respondent of firm 10 states: *What annoys me is the lack of environmental knowledge at the municipality level. They tend to apply environmental rules and procedures without paying attention to the firm's situation. By contrast, the people at the provincial level are much more knowledgeable and prepared to discuss environmental problems and find solutions.* Also the environmental coordinator of firm 8 perceives a lack of expertise at the municipality level: *Civil servants often stress, in fact, the less important issues, due to lack of knowledge.* The respondent of firm 2 points this out by indicating that the municipality he is dealing with attributes equal importance to reduced use of lighting on the plant-site compared to emissions of ammonia, while the latter implies far more severe risks for public health. He proposes: *The municipality should make an overview of the most important environmental issues and include them in their environmental permits, instead of emphasizing minor aspects, which distract the attention from the really important pollution items.*

In general, the respondents emphasize the negative consequences of overly strict monitoring by lower government. This is perceived as a barrier to information exchange with the municipality. *If I have an environmental problem, I will not inform the municipality first, because they will directly fine us. However, I will inform the province, because they are willing to jointly find a way to solve the problem.* Interestingly, the respondents stress the negative managerial and/or financial consequences of governmental demands. *The government often just applies environmental rules and procedures, while neglecting the fact that we have to implement them in our business operations* (firm 9). Illustrative is the following experience in firm 10: *Recently, I discussed a draft environmental permit with local governmental officials and we jointly concluded that about 75% of the originally included requirements should be omitted, because they had little or no environmental importance.* Firms that deal with different provinces and municipalities, because of multiple plants, seem to have an advantage using their experience: *Civil servants tend to stress certain issues based on their own expertise. However, if I disagree with the requirements, I refer to other provinces and/or municipalities, which do not impose them. Over the past few years, this has proven to be a successful strategy* (firm 4). Another strategy is mentioned by the respondent of firm 2, who sometimes tries to get in contact with higher municipality officials, because they are more flexible. In conclusion, most respondents indicate having found satisfying solutions to environmental problems together with lower government. In this context, the environmental coordinator of firm 12 states: *If you stay in discussion with lower governmental officials and show respect for their point of view, they are also willing to listen to the firm's viewpoint.* In line with the second proposition, it stresses the importance of good contacts with lower government to overcome problems with environmental requirements that are imposed from a command-and-control governmental perspective (see also Section 5.4.1).

Environmental covenants

Many respondents stressed that their participation in various covenants is, in fact, not voluntary but enforced, since it reduces costs to comply with regulations (firm 2, 3, 4, 5, 7, and 8). The respondent of firm 3 refers to this as follows: *We signed covenants, because we have to comply with the included environmental targets, such as on reduction of energy usage or packaging, sooner or later, anyway* (firm 3). This is also felt by the respondent of firm 7, who states that, if he did not participate, the municipality would ask him to report on energy consumption and packaging anyway. Remarkably different opinions exist on the effectiveness of covenants. Table 6.2 summarizes the positively and negatively evaluated issues, which are discussed below.

- Meat processing firms 7 and 8 both participate in the Long-Term Agreement Energy (see also Section 3.2). They stress the administrative consequences of the agreement, while it delivers no business value to them. This can be understood from the fact that energy reduction is already a very important issue to deal with in order to stay competitive.
- A medium-sized brewery (firm 1) is positive about the Long-Term Agreement Energy, because of the valuable feedback from SenterNovem, which is a governmental agency specializing in sustainable development and innovation, which evaluates environmental performance.
- A large brewery (firm 2) has signed the Energy Efficiency Benchmarking (see also Section 3.2). However, the respondent would prefer energy targets to be included in environmental permits, since the breweries now report collectively and anonymously via their branch-organization (for confidentiality reasons). According to the interviewee, the main drawback of this procedure is that the less efficient breweries are covered by the more efficient ones.
- The corporate environmental coordinator of a large dairy processor (firm 4) is, in general, positive about the firm's participation in various covenants (e.g. Long-Term Agreement Energy and Packaging Covenant), because it forced the individual plants to take into account important developments regarding (future) environmental regulation.
- The interviewee of a medium-sized dairy processor (firm 3) finds participation in covenants helpful, because he uses the results of the sector as a benchmark for the firm's own performance. With respect to the overall effectiveness of covenants, he is afraid that many

Table 6.2. Assessment of environmental covenants.

Positively evaluated	Negatively evaluated
Valuable feedback	Enforced regulation rather than voluntary
Performance benchmark	Collective environmental reporting
Attention for (future) environmental regulation	Administrative load
	Focus on requirements rather than attention for the environment

firms use environmental covenants *pro forma* rather than to really increase their attention for the environment on a structural basis. Referring to the first proposition (see Section 5.4.1), it might imply that participation in environmental covenants is not necessarily contributing to the development of environmental management capabilities.

Information on (future) environmental regulation

The respondents of the medium-sized and large food and beverage firms stay informed about (future) environmental regulations via different information sources, such as branch-organizations, professional journals, and the governmental intermediary agency InfoMil, which provides information on several environmental regulations. An often mentioned issue with respect to future regulations is the integration of EU regulation in the Dutch environmental regulations. Illustrative of the mainstream of comments made, is the response from firm 12: *Dutch environmental regulation has to be replaced by EU regulation sooner or later (e.g. regulations on waste water disposal and air emission), which may imply extra costs for new environmental investments to comply with EU requirements.* He is optimistic about future developments though, because he expects that problems can be collectively dealt with in cooperation with government. The respondents referred remarkably less frequently to EU regulation in 2007 compared to 2005. The interviewee of firm 13 confirms this: *I also have the impression that EU environmental regulation is less discussed as a problem these days compared to a few years ago.* This may be due to the fact that government is indeed willing to search collectively for solutions to implement EU regulations efficiently and/or because branch-organizations and firms themselves have taken sufficient measures to anticipate developments at the EU level.

Central versus lower governmental environmental policies

If a discrepancy is perceived between central and lower government, the respondents often relate this to a perceived lack of cooperation between governmental bodies and/or insufficient environmental knowledge at lower government (firm 2, 3, 4, 7, 10). For example, firm 8 participates in an environmental covenant, which was agreed with central government and should be monitored by lower government. However, it appeared that this was not adequately communicated and, therefore, the municipality lacked the knowledge needed for a proper evaluation. Furthermore, the interviewee of a large brewery (firm 2) says that he often has to inform the civil servants about developments in environmental regulations at the EU level.

6.2.2 Chain and network actors

A distinction can be made between vertical (e.g. with buyers and/or suppliers) and horizontal (e.g. with branch-organizations and/or competitors) environmental information exchanges. Vertical environmental cooperation with buyers and/or suppliers is carried out only by a few firms, such as a medium-sized brewery (firm 1) which cooperates with farmers of biologically cultivated products. Another example is firm 6, which aims to find solutions to reduce packaging material, using big bags and containers that are developed in collaboration with the buyers. Furthermore, an interesting situation was discussed with the respondent of a medium-sized bakery (firm 5), since it appeared that bakery factories often cooperate in consortia, which take care of collective buying of bakery ingredients and the supply to individual factories. Participation in these consortia is subject to several conditions, such as the guarantee of food quality and safety standards. *Environmental care is not an important issue yet, although the large bakery factories in particular might be increasingly confronted with it* (firm 5). Nevertheless, the respondent mentions several environmental initiatives, such as collective disposal of waste and reduction of packaging material in cooperation with suppliers.

The main barrier to environmental cooperation at the chain level is a lack of environmental interests downstream, including retailers and consumers. Illustrative is the remark by the respondent of a medium-sized brewery (firm 1): *Retail is not interested in our environmentally friendly produced beer, because it is only a minor share in their overall turnover.* In addition, it is said that even if customers are interested in environmental issues, guarantees of food safety and quality issues still prevail: *Although our customers (e.g. large fast food firms) pay attention to the environment, their audits mainly stress food safety and quality aspects* (firm 11). Overall, the most important reason for vertical information exchanges appears to be administrative requirements following on from the Packaging Covenant. The environmental coordinator of firm 3 remarks: *It is particularly difficult to make agreements on the use of the same packaging for a longer period with foreign suppliers.* Since the Packaging Covenant ended in January 2007, the interest in packaging issues has waned in some firms: *Now we only keep administration records on packaging of products that are sold to the consumer market, because our customers want us to provide this information* (firm 9).

None of the firms cooperate on environmental issues with competitors. However, environmental information exchange frequently takes place via the branch-organizations. They appear to play an important role in stimulating firms to share environmental experiences. Since the exchanged environmental information is generic in terms of environmental measures taken and activities, competition forms no barrier, according to most respondents. The breweries (firm 1 and 2) and dairy processors (firm 3 and 4) in particular were positive about the role of their branch-organization. As expected, the branch-organizations appear also an important linkage between government and firms: they try to influence (future and/or proposed) environmental regulations both at the national and EU level and conclude public-private environmental agreements. The only respondents who point to a lack of support from a branch-organization are from the two slaughterhouses (firm 7 and 8). Table 6.3 summarizes the discussed reasons for and barriers to vertical and horizontal environmental information exchange in the Dutch food and beverage industry.

6.2.3 Societal groups

Two important societal groups are environmental organizations and local inhabitants. The environmental coordinator of firm 12 maintains contact with various environmental organizations: *Discussions with them provide us with useful insights and even solutions to environmental problems, but it does not work when they show an aggressive attitude towards the firm.* The respondent of firm 1 states: *We have been working together with environmental organizations on projects, which resulted in interesting alternatives to reduce energy usage.* However, none of the firms perceives strong pressure from environmental organizations to clean up. By contrast, local inhabitants are exerting more pressure and are therefore regarded as important. Different strategies are used to safeguard a good relationship with the local community. The most important one is responding well to the complaints by getting in touch with the people personally. As the respondent of firm 10 states: *It is very important that people feel free to complain to our firm directly, because otherwise they might go straight to government, which would probably cause us a lot of trouble with respect to environmental permits.* In line with this observation, several firms regularly inform local inhabitants about their environmental activities via information bulletins and, in some cases, open-days at the plant-site. Other strategies include regular meetings with people from the local community to provide insight into environmental activities and to discuss environmental issues (firm 1, 2, 10, and 12).

Table 6.3. Horizontal and vertical environmental information exchange in chains and networks by Dutch food and beverage firms.

Vertically	**Horizontally**
With suppliers to gather information needed to comply with covenants	Via branch-organizations with competitors to exchange environmental experiences
Main barrier: Lack of environmental interest downstream in the chain (e.g. retail and consumers)	*Main barrier:* Strong competition and/or a lack of a supportive branch-organization

6.3 Environmental capabilities

6.3.1 Motives for strategic commitment

Among the most important reasons to pay attention to the environment are regulative requirements and cost savings, such as less organic waste and/or energy usage. Illustrative is a recent development at a medium-sized dairy processor (firm 3), where the respondent is appointed by the board as environmental coordinator in response to growing pressure from the province and the local community to reduce pollution (in terms of noise, smell, and waste water). There was clearly an urgent need to clean up, according to the interviewee, not only to meet stakeholder expectations, but also to avoid more costly lawsuits. Although environmental complaints still provide an important motive for further improvement of environmental performance, cost-savings have become a strong driver over the past few years as well. As the environmental coordinator states: *Most of our environmental measures generate financial benefits, which is, I believe, an important reason for my ongoing appointment* (firm 3). Also at other firms, including the slaughterhouses (firm 7 and 8) in particular, cost savings are important drivers for environmental care.

Another strategic motive is found at a medium-sized brewery (firm 1), which increased its attention for the environment a few years ago, as part of a sustainability strategy to maintain market share. The consequence is a higher price paid for raw materials (about 10%), which is reflected in a higher sales price than the competitors. With respect to the benefits of the firm's sustainability strategy, the respondent states: *Due to the decline in the beer market as a whole, it is difficult to point out the economic benefits of our sustainability strategy. However, I am convinced that we would be in a worse position otherwise.* Also in several other firms, environmental commitment is associated with a firm strategy that stimulates proactive attention for the environment. For instance, the interviewee in firm 11 indicates that it fitted their innovative business strategy: *Environmental care contributes to our corporate image, because we also want to show our innovative reputation in environmental management.* Environmental excellence is also included in the corporate strategy of a large dairy processor (firm 4), since formal environmental responsibilities are laid at the Board of Director's door. The firm's respondent emphasizes the importance of this in clearly demonstrating to the external stakeholders the environmental commitment of the firm. Table 6.4 summarizes the main motives found for strategic commitment, divided into low-cost production and product differentiation strategies.

6.3.2 Organizational structure and responsibilities

The respondents of two medium-sized firms are also members of the Board of Directors (firm 1 and 5). This implies that they have a direct impact on environmental issues. By contrast, the respondents of almost all the other firms have a supportive (or staff) function (firm 2, 3, 4, 8, 9, 10, 11, 12, and 13). The corporate environmental coordinator of a large dairy

Table 6.4. Reasons for strategic commitment to the environment.

Low-cost production strategy	Differentiation strategy
Compliance with environmental regulations	Safeguarding market position
Meeting pressure from the local community	Reputation enhancement
Cost-savings by reducing pollution	

processor (firm 4) states: *The formal environmental responsibilities lie at the door of the Board of Directors, as well as at the plant managers' level. My task is to communicate environmental issues in the organization and to provide support to the plant-sites.* Hence, in most firms the line-responsibility for the environment is in the hands of the plant managers and/or the head of different departments (e.g. production and logistics). Despite the fact that formal decision-making power is lacking, the majority of the interviewees indicated that they have a strong impact in case of environmental issues. This is achieved through close contacts with (members of) the board of directors on a monthly (firm 3 and 4) or weekly basis (firm 2 and 11). As the respondent of firm 2 states: *Every Monday morning I evaluate environmental issues with my director.*

6.3.3 Embeddedness in the organization

All respondents indicate that different people in the organization are consulted to deal with environmental issues. Most respondents of the medium-sized firms point out that it is easy to get in touch with employees, managers, and other experts in the organization, as the firm has a flat organizational structure (firm 5, 7, 9, and 13). The respondents in another medium-sized (firm 3) as well as in the large firms perceive, however, more difficulties in keeping people involved. The impression emerges that this is due to the fact that managers have other managerial priorities, constrained by a limited financial budget. Illustrative is the following remark made by the interviewee of firm 2: *Many environmental measures save money, but if they don't, as in the case of noise reduction, I have far more problems convincing people to take care of it, despite the legal urgency to do so.* Increasing environmental awareness appears to be important for enhancing internal cooperation. As in firm 3, where the environmental coordinator stresses that active communication of environmental information is essential: *If I don't keep drawing attention to the environment, people will quickly forget about it.* He therefore issues an environmental news bulletin for employees four times a year. In firm 12, environmental information is shared online via the firm's intranet. Furthermore, multiple firms implement environmental training and education programs (firm 2, 3, 4, 9, 11, and 12).

Despite the fact that people from different departments are involved, it is noticeable that the respondents keep closer contacts with production and logistics than with marketing, sales, and R&D[13]. Typically, the environmental team in firm 9 consists of the head of technical service, the plant manager, and the environmental coordinator. On the one hand, this bias might be related to the strategic position of the respondents: several of them indicate that marketing, sales, and/or R&D issues are not of (formal) concern to them (firm 2, 3, 4, 6, and 10). Hence, it does not necessarily imply that top management is not delegating environmental responsibilities to these departments. On the other hand, a lack of attention for the environment in R&D is confirmed by two respondents (firm 2 and 11). The environmental coordinator in firm 11 even senses an adverse attitude towards environmental targets by the R&D department: *I am afraid that they will interpret it as a creativity killer.* Nevertheless, the firm has strategic plans to involve R&D more closely in environmental management, starting with environmental training and education organized by the respondent. In conclusion, the attention for environmental issues in the R&D processes is limited, despite the fact that it is essential to develop cradle-to-cradle environmental capabilities needed for environmentally friendly product (re)design. Referring forward, this is strikingly in concordance with the results of the quantitative analysis (see Section 7.3.1 and 7.4.1).

6.3.4 Integration of care systems

Many respondents indicate that the environment is jointly considered with food quality and safety issues (firm 1, 3, 4, 7, 8, 10, and 11). The integration of the two is taking place especially at the operational level in terms of documentation and auditing (firm 4, 7, 8, and 10). Illustrative is the way in which the interviewee of firm 10 introduced and documented environmental issues. He analyzed, in cooperation with other people (e.g. quality manager, head of operators, and the plant manager), the risk of pollution for the different production lines. This resulted in a list of operational measures, such as daily checks on chemical storage and more frequent maintenance of equipment to save energy. Next, these measures were included in the existing handbooks, used by operators to safeguard food quality and safety, as well as health and safety. Looking at the joint audits, they consist of environmental and food quality and safety issues, but also of other checks, such as on maintenance of machines and safe working conditions (firm 1, 4, 9, and 10). For instance, both a medium-sized brewery (firm 1) and meat processor (firm 9) audit their factory once every 4-6 weeks to evaluate these issues.

6.3.5 Information processing

The respondents obtain information on environmental performance from different plants and/or departments through internal environmental reporting: it is carried out on a yearly (firm 8), monthly (firm 1 and 9), or weekly (firm 4 and 7) basis. Reporting formats are developed to ensure uniform documentation. For example, firm 12 uses the Global Reporting Initiative guidelines as a basis (see also Section 3.1), while firm 7 and 8 keep a spreadsheet of different environmental issues, such as energy consumption, water usage, etc. Environmental

---

[13] Marketing and R&D are lacking in medium-sized firms 5 and 7.

and other internal reporting obligations are sometimes integrated. In a large dairy processor (firm 4), plant managers have to develop medium-term business plans (time horizon of 3-5 years), which should consider environmental targets as well. The respondent stresses that these plans are very useful for anticipating business activities from an environmental point of view. On the whole, the respondents indicate that there is very limited or no insight into the costs of environmental investments, because they are not registered and reported as such. Instead, they are integrated in other investments, such as those to safeguard food quality and safety and/or maintenance activities (firm 1, 2, 7, and 8).

6.4 Concluding remarks

The overall impression that emerges from the semi-structured interviews is that the adoption of environmental capabilities is strongly associated with managerial control of manufacturing processes (e.g. less energy usage and waste water production). This is reflected in the involvement of people from production and logistics, while marketing, sales, and R&D is consulted to a far lesser extent. These departments focus on satisfying customer demands in terms of product characteristics, such as price, volume, taste, and design. A limited environmental perceived interest of buyers and consumers might explain the lack of environmental interest in these departments. It has important implications in terms of the limited development of chain-oriented and cradle-to-cradle capabilities, which are necessary to reduce environmental impact from a product-life-cycle point-of-view. Among the most important drivers for the adoption of environmental management capabilities are governmental environmental regulations and pressure from the local community. In addition, the majority of the interviewees indicate financial benefits from improved production efficiency as an essential driver as well.

Table 6.5 shows the confrontation of the propositions with the empirical findings. The propositions focus on different issues that are proposed to be positively related to the adoption of environmental management capabilities. The first proposition (P1) concentrates on environmental covenants. It is confirmed by the interviews, since it is found that valuable environmental feedback is obtained from participation in environmental covenants. However, it should also be noted that criticism is expressed by the respondents, such as collective reporting and a focus on environmental requirements rather than attention for the environment (see Table 6.2). It implies that environmental covenants do not always stimulate firms to 'green' their business. The second proposition (P2) proposes a positive relationship between the contact intensity with lower government and the attention for environmental management capabilities. It is partly confirmed based on the fact that the majority of the interviewees perceives a lack of environmental knowledge at the municipality level, while contacts at the provincial level would be used to discuss environmental problems and find solutions together with the provincial government. In line with the third proposition (P3), the results confirm the role of an intermediary, such as a branch-organization, to support and stimulate firms to pay attention to the environment. Important activities that are mentioned include facilitation of sharing environmental experiences between firms and involvement in the settlement of

Table 6.5. Empirical assessment of the propositions.

Focus of the proposition		Results
(P1)	Environmental voluntary agreements	++
(P2)	Contact intensity with lower government	+
(P3)	Environmental support from an intermediary	+++
(P4)	Pressure from chain actors	+
(P5)	Prospector strategy and cradle-to-cradle capabilities	++
(P6)	Top-management commitment	+++
(P7)	Perceived impact of an environmental manager	+++
(P8)	Horizontal and vertical communication	+++
(P9)	Availability of (other) care systems	++

The following criteria are applied: +++ = confirmed by all respondents; ++ = confirmed by a majority of the respondents; + = confirmed by a minority of the respondents.

public-private environmental voluntary agreements. The fourth proposition (P4) is confirmed to the extent that several interviewees are working together with buyers and suppliers to reduce the use of packaging materials and guarantee environmentally friendly product ingredients. However, the interviewees perceive in general limited or no environmental interest among retail chains and consumers.

The fifth proposition (P5) suggests that prospector firms have cradle-to-cradle environmental management capabilities needed for environmental product (re)design. Although several respondents indicate that showing environmental excellence is part of their business strategy, the attention for cradle-to-cradle capabilities is very limited. The empirical findings are in concordance with the sixth proposition (P6) which states that top-management commitment is essential for the adoption of environmental management capabilities. It appears necessary to force people in the organization to pay attention to the environment. Additionally, the appointment of a board member with environmental responsibilities is stressed as beneficial in relationships with government in order to influence future or proposed environmental regulations. Also the seventh proposition (P7) is confirmed by the empirical data, since the interviewees indicate that their role as environmental manager is of central importance in maintaining environmental awareness in the different departments. In accordance with proposition eight (P8), the establishment of environmental communication channels is indicated as very helpful to obtain environmental feedback from employees and to inform higher level management. Finally, empirical evidence is found for proposition nine (P9), since existing food quality and safety procedures are used to include environmental issues as well. It is emphasized, though, that food quality and safety issues clearly get a higher managerial priority than environmental issues.

7. Survey results

This chapter presents the results of the survey data gathered in 2002 and 2005. Section 7.1 discusses the general characteristics of the study samples, the response rates, as well as the representativeness for the Dutch food and beverage industry as a whole. Furthermore, the background of the respondents is evaluated and a non-response analysis is carried out. Section 7.2 focuses on the validity and reliability of the constructs. Section 7.3 presents the results of the survey carried out in 2002. It starts with assessing mean values for the micro, small, medium-sized, and large firms. The Spearman rank correlations are evaluated preceding a linear regression analyses to reveal the most significant predictors of attention for environmental management capabilities. Section 7.4 includes the results of the survey conducted in 2005 among the medium-sized and large firms. It evaluates the mean values, Spearman rank correlations, and findings of the linear regression analyses. Furthermore, it elaborates on the results of the cluster analysis, which is carried out to get more insight into the joint impact of the business network and firm characteristics on the development of environmental management capabilities. Section 7.5 focuses on the longitudinal analysis comprising medium-sized and large firms only. It evaluates changes business network impacts between 2002 and 2005. It discusses the Spearman rank correlations and the results of the regression analysis. Finally, Section 7.6 provides the concluding remarks with respect to this chapter and it links the empirical findings to the propositions.

7.1 Study samples

7.1.1 Response

Table 7.1 shows the mailing and response of the survey carried out in 2002.

Table 7.1. Mailing and response in 2002.

	Number of firms	
Mailed firms		
Total sample (firms with ≥ 5 employees)	2,627	
Of which firms with ≥ 50 employees	356	
Response		
Received questionnaires	590	
Non-usable questionnaires (blank, incomplete, etc.)	98	
Total usable questionnaires (response rate)	492	(19%)
Of which firms with ≥ 50 employees (response rate)	106	(30%)

The survey included 2,627 Dutch food and beverage firms in total, of which 492 usable questionnaires were received: this equals an effective response of 19%. No second mailing was carried out. As regards medium-sized and large firms (≥ 50 employees), 106 received questionnaires could be used for the analyses: this corresponds with an effective response rate of 30%, which is higher than the overall response. Apparently, medium-sized and large were more willing to cooperate than the micro and small firms. This might be due to a greater urgency to pay attention to the environment in large firms and/or because they were more likely to have appointed an environmentally responsible manager, who could take care of the questionnaire (see also Section 7.1.2). Furthermore, it could be that large firms have more environmental information available to answer the questionnaire, because of environmental reporting obligations.

As already stated in Section 5.5, only firms with 50 or more employees are included in 2005. This was done to be able to analyze the importance of firm strategy and enabling environmental capabilities in more detail. Table 7.2 shows that 426 addresses of medium-sized and large Dutch food and beverage firms are included in the mailing in 2005. Notably, the number of medium-sized and large Dutch food and beverage firms increased between 2002 and 2005, according to the database obtained from the Dutch Chamber of Commerce (compare Table 6.1). Nine addresses turned out to be wrong (e.g. firm had moved). Two mailings are carried out, resulting in 100 completed and usable questionnaires: this equals an effective response rate of 24%. After the first mailing, the reasons for non-response are investigated randomly at 150 firms. The main reasons for withholding cooperation appear to be a lack of time and/or a corporate policy that restricted respondents from participation, because of confidentiality reasons. This is in line with reasons for non-response found by Ghobadian *et al.* (1998). In addition, medium-sized firms in particular indicated that they lacked a manager with environmental responsibilities and/or sufficient environmental knowledge to complete the survey questionnaire. In the other cases, the name of the environmentally responsible manager is requested and included in the second mailing.

Table 7.2. Mailing and response in 2005.

	Number of firms	
Mailed firms		
Total sample (firms with ≥ 50 employees)	426	
Wrong addresses	9	
Total sample	417	
Response		
After the first mailing	65	
After the second mailing	35	
Total usable questionnaires (response rate)	100	(24%)

The response rate of the medium-sized and large firms is lower in 2005 compared to 2002. The questionnaire used in 2002 included an evaluation of administrative consequences of governmental regulation as well, which is carried out on behalf of the Ministry of Housing, Spatial Planning, and the Environment (see Section 5.1). The perceived urgency to take care of the administrative burden of governmental regulations is expected to have stimulated participation and, therefore, increased the response rate. Furthermore, the lower response rate in 2005 might be associated with a lack of insight into the requested strategic and environmental organizational characteristics (e.g. firm strategy and top-management commitment) or hesitation to share this information. On the whole, the yielded response rates are favorable compared to other studies that employed environmental surveys, see Section 2.2 and 2.3 for an overview (Table 2.1 and 2.2).

7.1.2 Position of the respondents

Table 7.3 shows the position of the respondents. The category 'others' consists of very diverse functions, including plant manager, technical service, marketing, and external affairs manager. A distinction is made between micro (5-10 employees) and small (10-50) versus medium (50-250) and large (> 250 employees) firms in 2002. Table 7.3 demonstrates that most respondents in the micro and small food and beverage firms are owner and/or CEO, implying that they take care of environmental issues themselves. By contrast, the position of the respondents in the medium-sized and large firms indicates that environmental responsibilities are frequently delegated to other people in the organization, such as an environmental coordinator or quality manager, as well as combinations of these functions. The fact that relatively many respondents have multiple responsibilities is in line with the results from other studies (Sharma, 2001; Banerjee, 2002; Sroufe, 2003; Clemens and Douglas, 2006). The changes between 2002 and 2005 show an interesting increase in environmental coordinators and quality managers in the medium-sized and large firms. The participation of CEOs and board members also

Table 7.3. Position of the respondents.

	Micro and small firms	Medium-sized and large firms	
	2002 (N=373)	2002 (N=106)	2005 (N=100)
Environmental coordinator	1%	13%	18%
Quality manager	5%	12%	24%
Owner and/or CEO	38%	2%	5%
Board-member	8%	1%	12%
Multiple functions	37%	51%	20%
Others	11%	21%	21%
Total	100%	100%	100%

increased. This might reflect growing attention for the environment from a strategic point of view. Furthermore, top management involvement in the survey in 2005 could be due to the focus of the questionnaire, which required knowledge on firm strategy and environmental organizational characteristics. It is assumed that the respondents were sufficiently informed to complete the questionnaires and provide reliable information. The fact that all returned questionnaires in 2005 could be used for the analyses supports this assumption, since an adequate questionnaire and the use of knowledgeable respondents reduces the chance of missing data (Stocké, 2006).

The different positions of the respondents may have implications with respect to their answers on the questions, such as the perceived quality of environmental information provided by lower government and the perceived environmental commitment of the board. The answers on all research variables are therefore checked for the respondent's position. No significant differences are found between the different functions in the micro and small firms in 2002. By contrast, the environmental coordinators in the medium-sized and large firms indicate a significantly higher level of environmental management capabilities compared to the other respondents in 2002 (P<0.05). This was to be expected, since it is their job to stimulate the development of environmental management capabilities. The environmental coordinators also indicate a stronger perceived impact on operational environmental activities compared to the quality managers in 2005 (P<0.05). This result can be logically explained by the focus of environmental coordinators on the planning and implementation of environmental activities. No other significant differences in mean values on the research variables are found between the positions of the respondent in 2002 and 2005.

7.1.3 Non-response analysis

In order to detect non-response bias, it is proposed that late respondents have the same characteristics as non-respondents. The data of 2002 are gathered in one batch, which makes it hard to distinguish between two groups. A non-response analysis is therefore carried out for the data of 2005 only. A distinction is made between respondents that answered after the first (N=65) and second (N=35) mailing. All research variables are included in the analysis. The only two significant differences appear to be a higher average level of environmental pollution and a greater average level of impact of employees on environmental management found for the second compared to the first batch (P<0.05). More particularly, higher levels are indicated for water and air pollution, as well as noise releases (P<0.05). Hence, respondents from firms with a higher level of environmental impact might have been slower in replying. It appears also that significantly fewer quality managers and more people in the category of other functions (e.g. technical service, marketing, and manager external affairs) are included in the second compared to the first batch (P<0.01). The mailing was directed at the 'environmentally responsible manager' in general. The name of this manager was requested by telephone at 150 firms for the second mailing (see Section 7.1.1). This may have enhanced the participation of people in functions like technical service and marketing in particular, because they might not be directly associated with environmental responsibilities. The fact that the respondents included in the second compared to the first batch indicated a higher level of environmental impact might reflect a greater need to clean up, but also a lack of good insight into the environmental pressures, because environmental issues will not be of major interest to the people from technical services and marketing.

7.1.4 Company size and sectors included

The following distribution of the study sample gathered in 2002 is found (N=492): micro (41%), small (37%), medium-sized (17%), and large (5%) firms. For reasons of comparison with 2005, the distribution of medium-sized and large firms only in the sample of 2002 is 76% medium-sized and 24% large firms. The study sample of 2005 consists of 79% medium-sized and 21% large firms (N=100). A visual check did not indicate severe deviations from the Dutch food and beverage industry as a whole. Table 7.4 shows the included sectors. The category 'others' comprises processors of sugar, cocoa and coffee (among others).

Some striking results can be observed in Table 7.4. The total Dutch food and beverage industry and both the study samples consist of relatively many firms from the bakery and the meat sector. It can be noted that the first sector, which produces bakery and pastry products, is especially well represented in the sample of small and micro firms. Despite some minor differences, a statistical analysis indicates no significant differences between the sectors as

Table 7.4. Food and beverage sectors.

	Small and micro firms		Medium-sized and large firms			
	2002		2002		2005	
	Total population (N=2,290)	Study sample (N=373)	Total population (N=356)	Study sample (N=106)	Total population (N=426)	Study sample (N=100)
Meat	14%	14%	26%	24%	20%	20%
Bakery	66%	62%	23%	16%	26%	26%
Vegetables and fruit	3%	5%	9%	10%	8%	11%
Animal feed	4%	5%	8%	8%	6%	3%
Beverages	1%	1%	6%	8%	6%	8%
Dairy	2%	2%	4%	9%	11%	10%
Grain mill and starch	1%	3%	3%	5%	3%	6%
Others	9%	8%	21%	20%	20%	16%
Total	100%	100%	100%	100%	100%	100%

included in the study samples compared to the total population. Also the distribution over the different sectors does not differ significantly between the study samples and the total population ($\chi^2 > 0.10$). In conclusion, all samples are representative for the Dutch food and beverage industry as a whole.

7.2 Validity and reliability

The validity and reliability of the reflective and formative constructs was checked (see also Section 5.6). The content validity is established for both types of research variables by a careful literature review (see Chapter 2 and 4). Furthermore, the convergent, discriminant and nominal validity are statistically assessed for the reflective constructs. Convergent validity is evaluated by means of factorial data reduction: the variables comprising the constructs were jointly entered in a factor analysis (Principle component; Varimax rotation) in 2002 and 2005. Appendix F shows the results: all factor loadings are greater than 0.60, which is satisfactory (Field 2003). The discriminant validity is supported by the Spearman correlations, which are all below 0.80 (see Table 7.13, 7.14, 7.23, and 7.25). Nominal validity is evaluated based on the results of the regression analyses (see Table 7.15, 7.16, 7.21, and 7.26). The significance of the constructs in the regression models indicates a sound nominal validity.

The reliability of the formative constructs is tested during the semi-structure interviews by focusing on the research variables. From the discussions, it is concluded that the respondents had a sufficiently uniform interpretation of the included constructs. The reliability of the reflective constructs is statistically tested with Cronbach α, which measures the internal consistency of the items thought to reflect a single construct. The results are included in Appendix F: all constructs have a sufficiently high Cronbach α, exceeding the lower limit of 0.70 in almost all cases, except for some which are still well above the lower limit of 0.60 for explorative research (Hair *et al.*, 1998).

7.3 Analyses of the 2002 survey data

This section presents the results of the first survey carried out. The data are gathered in 2002 in micro, small, medium-sized, and large Dutch food and beverage firms.

7.3.1 Environmental management capabilities, pollution and budget

Table 7.5 includes the percentage of firms that implemented different elements of the environmental management capabilities. A distinction is made between strategic, operational, as well as chain and cradle-to-cradle environmental management capabilities in Table 7.5. In general, more attention is paid to the strategic (no. 1 and 2) and operational environmental management capabilities (no. 3 till 10) by the medium-sized and large compared to the micro and small firms. Less straightforward results are found for the chain and cradle-to-cradle environmental management capabilities (no. 11 till 13). Overall, they are less available in all

Table 7.5. Percentage of firms (%) paying attention to environmental management capabilities in micro (5-10 employees), small (10-50), medium-sized (50-250), and large (>250 employees) food and beverage firms in 2002 (N=492)[1].

	Micro (N=196)		Small (N=177)		Medium (N=81)	Large (N=26)
Strategic capabilities						
1. Formulation of an environmental strategy	12	—	22	—	54	69
2. Formulation of an environmental action program	4	—	13	—	45	65
Operational capabilities						
3. Info. collection to check environmental emissions	6	—	26	—	67	77
4. Regular checks on environmental impact	7	—	23	—	58	69
5. Internal environmental information processing	7	—	15	—	49	69
6. Info. collection to evaluate environmental measures	4	—	13	—	42	62
7. Info. collection for internal environmental care	16	—	26	—	59	59
8. Environmental audit	5	—	15	—	61	58
9. Environmental database	3	—	9	—	41	58
10. Environmental training of employees	16		15	—	33	46
Chain and cradle-to-cradle capabilities						
11. Info. collection for chain-oriented environ. care	4	—	10	—	19	27
12. Info. collection to exchange with buyers/suppliers	13		18		11	23
13. Info. collection for environ. product (re)design	8		8		9	8

[1] Significant differences between size categories are indicated by — (P < 0.05); the box indicates that no significant differences were found within the medium-sized and large groups.

firms compared to the other environmental management capabilities. Only the collection of information for chain-oriented environmental care (no. 11) is more frequently carried out by the medium-sized and large compared to the micro and small firms. This can be related to their participation in public-private environmental voluntary agreements, such as the Packaging Covenant. However, the micro and small firms pay, in comparison with the other elements listed in Table 7.5, rather substantive attention to environmental information collection (no. 12). This might be related to large buyers' requests. Also large suppliers could play an important role in this, if they search for environmental cooperation to reduce packaging waste (like the large supplier of bakery ingredients interviewed in Section 6.2.2). Environmentally responsible product (re)design (no. 13) is implemented by very few firms. Looking at the sector level, no obvious differences in the implementation of environmental product (re)design were found, with the exception of the grain mill sector (N=15). Three small and two medium-sized grain mill processors appear to pay attention to it (equaling 33%). Although the sample of grain mill processors is small, their remarkable interest in environmental product (re)design could be related to the increased attention for genetically modified ingredients.

Figure 7.1 shows the indicated importance of different pollution issues in the micro and small versus medium-sized and large firms. A Mann-Whitney test indicates that the higher importance attributed to pollution issues in the medium-sized and large compared to the micro and small firms is significant (P<0.01). Overall, Figure 7.1 shows, nevertheless, that the respondents assess their environmental impacts as moderate at most. A similar pattern of most important issues is observed for the two groups of firms. Water pollution and waste production are among the most frequently indicated pollution issues. This is in line with the figures found for the Dutch food and beverage industry as a whole (see Section 3.3.2) and the results of the semi-structured interviews (see Table 6.1), although the interviewees also indicated a high level of energy usage, which was not included in the survey. Other frequently indicated pollution issues that can be observed in Figure 7.1 are noise and smell releases. The results of the interviews suggest that the urgency to take care of this depends strongly on the presence of a local community in the neighborhood of the firm that may complain about local noise and smell hindrance.

Table 7.6 shows the mean values on the availability of environmental management capabilities (i.e. the sum of the elements listed in Table 7.5 and rescaled on a 0-5 point scale), the pollution level, and the environmental budget. It demonstrates that significantly more environmental management capabilities are available in the medium-sized and large compared to the micro

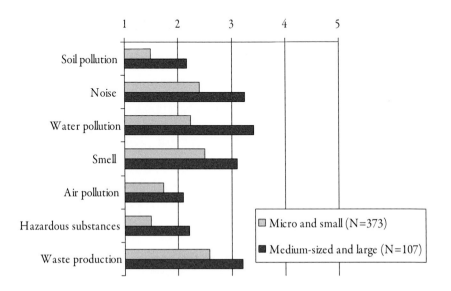

Figure 7.1. Environmental pollution impacts (1=Not important; 5=Very important).

Table 7.6. Environmental management capabilities, pollution level, and environmental budget.

	Micro (N=196)	Small (N=177)	Medium-sized (N=81)	Large (N=26)
Level of environ. capabilities	0.40 (0.56)[a,b]	0.80 (0.84)[a,c]	2.08 (1.31)[a,d]	2.69 (1.39)[b,c,d]
Pollution level	2.36 (0.98)[a]	2.58 (0.96)[b]	3.24 (0.82)[a,b,c]	3.25 (1.02)[a,b,c]
Environmental budget	1.69 (0.99)[a,b]	2.06 (1.01)[a,b]	2.58 (0.84)[a,e]	3.12 (0.78)[b,e]

Significant differences are indicated with [a, b, c] $P < 0.01$; [d] $P < 0.05$; [e] $P < 0.10$.

and small firms. There are four reasons for this. First, the medium-sized and large firms have a significantly stronger impact on the environment, which makes it more necessary to clean up (see also Figure 7.1). Second, the medium-sized and large firms have more resources to invest in environmental issues. This can be seen in Table 7.6 as the environmental budget is significantly higher in medium-sized and large compared to micro and small firms. Third, the medium-sized and large firms have a greater urge to implement environmental management capabilities, because their scale requires more formal planning of environmental activities. Fourth, the development of environmental management capabilities could be related to stronger stakeholder influences to clean up (e.g. from government or local inhabitants) because a large size attracts stronger stakeholder interests.

7.3.2 Business network characteristics

Table 7.7 shows the mean values on different business network impacts. As expected it indicates that government is perceived as the most influential stakeholder, followed by local inhabitants. Buyers are also associated with impacts on the firm, although their influence is felt to be limited. The lowest impact is measured for competitors. This is in line with the interviews, since it reflects the lack of competition on environmental issues indicated by the majority of the interviewees. Overall, perceived stakeholder impacts are stronger in the medium-sized and large compared to micro and small firms, although not all differences are significant.

Table 7.8 shows the frequency of contacts with lower governmental agencies, including municipalities, provinces, and water boards. In line with expectations, a higher contact frequency is measured for the medium-sized and large compared to the micro and small firms. This might be related to renewal of environmental permits (e.g. new environmental requirements) and mandatory governmental environmental reporting to lower government. Overall, most frequent contacts take place with the municipalities and water boards.

Table 7.9 shows that larger firms assess the quality of the governmental environmental information provided more positively. By contrast, mixed patterns are found for the relevancy

Table 7.7. The perceived impact of the business network on environmental management (1 = No impact; 5 = Very strong impact).

	Micro (N=196)	Small (N=177)	Medium-sized (N=81)	Large (N=26)
Government	3.39 (1.18)[a,e]	3.68 (1.07)[e]	4.03 (0.75)[a]	3.69 (0.93)
Local inhabitants	2.60 (1.24)[b]	2.65 (1.26)[c]	3.08 (1.12)[b,c]	3.13 (0.95)
Buyers	2.20 (1.19)[b,c,d]	2.61 (1.29)[b]	2.66 (1.10)[c]	2.96 (0.96)[d]
Suppliers	1.88 (0.99)	2.09 (1.08)	1.96 (0.72)	2.29 (0.96)
Environmental organizations	2.03 (1.18)	1.96 (1.07)	2.16 (1.03)	2.17 (0.87)
Competitors	1.48 (0.74)[a,b]	1.76 (0.85)[b]	1.88 (0.85)[a]	1.79 (0.51)

Significant differences are indicated with [a] $P < 0.01$; [b, c, d] $P < 0.05$; [e] $P < 0.10$.

Table 7.8. Frequency of contacts on environmental issues with lower governmental agencies (1 = never, 2 = yearly, 3 = twice a year, 4 = four times a year, 5 = monthly).

	Micro (N=196)	Small (N=177)	Medium-sized (N=81)	Large (N=26)
Municipalities	1.85 (0.75)[a,b]	2.21 (0.95)[a,b]	3.10 (0.96)[a]	3.31 (1.12)[b]
Provinces	1.23 (0.48)[a]	1.43 (0.85)[b]	2.08 (1.22)[a,b]	2.80 (1.50)[a,b]
Water boards	1.54 (0.78)[a,b]	1.91 (1.07)[a,b]	3.06 (1.25)[a]	3.56 (1.23)[b]

Significant differences are indicated with [a, b] $P < 0.01$.

of the different information sources. The micro, small, and medium-sized firms prefer a visit from the governmental civil servant, while no such preference is observed for large firms (although the differences are not significant; see Table 7.9). This can be explained by the availability of more resources in the large firms, namely time and people (e.g. an environmental coordinator) to pay a visit to the civil servant. Moreover, in the interviews it was indicated by some respondents that paying visits is important to try to persuade lower government to make changes in environmental permits or (other) regulative requirements. The micro firms attach the least importance to a fixed contact person, because their environmental impact will, in general, be so small that they do not benefit from it.

Table 7.9 indicates a higher level of perceived equality and dialogue of the governmental relationship in the medium-sized and large compared to the micro and small firms. Two

Table 7.9. Different aspects of the governmental relationship (1 = I completely disagree; 5 = I completely agree).

	Micro (N=196)	Small (N=177)	Medium-sized (N=81)	Large (N=26)
Quality of information provided				
Consistency of info. on permits	2.53 (1.18)	2.56 (1.05)	2.80 (0.99)	2.96 (1.02)
Clear environmental policy	2.76 (1.07)	2.71 (1.00)c	3.04 (0.91)	3.24 (1.05)c
Relevancy as information source				
Visit of governmental civil servant	2.83 (1.13)	3.15 (0.97)	3.11 (0.93)	2.68 (1.03)
Visit to governmental civil servant	2.48 (1.07)	2.60 (0.97)	2.48 (0.95)	2.70 (0.92)
Fixed contact person	2.38 (1.10)c,d	2.85 (1.08)c	2.86 (0.91)d	2.83 (0.96)
Equality and dialogue				
Enough say in environmental policy	2.10 (1.01)c	2.21 (0.95)	2.35 (0.92)	2.64 (0.81)c
Open dialogue on policy goals	2.24 (1.01)a,b	2.59 (1.08)a,b	3.14 (1.00)a	3.58 (0.81)b
Informal commun. on policy goals	2.42 (0.94)a	2.57 (1.09)b	3.03 (0.91)a,b	3.76 (0.72)a,b
Perceived congruence				
Contribution to business goals	2.52 (0.96)c	2.68 (0.91)	2.82 (0.77)c	2.84 (0.80)

Significant differences are indicated with a, b $P < 0.01$; c, d $P < 0.10$.

causes can be mentioned in this context. First, they have to deal with environmental permits that are especially focused on their situation. The imposed emission criteria and required environmental measures in the permits might be a cause for debate with lower government. For example, a firm might want to relax demands on noise reduction or waste water disposal. By contrast, micro and small firms will be confronted with generic environmental regulation, such as on emission of noise and smell. Second, medium-sized and large firms are expected to have more power to influence government. The interviews indicated, for instance, that these firms generally have good insight into environmental policy developments at the sector and the EU level. Furthermore, medium-sized and large firms might contribute to other governmental interests, such as local employment and industrial growth, which could contribute to a less hierarchical relationship with lower governmental bodies.

Table 7.9 further demonstrates that the perceived level of congruence of governmental policy and own business goals is rather low. As expected, the lowest values are measured in the micro and small firms, because they want to focus on their core business (e.g. meat processing and baking bread). As stated in Section 6.2.1, several interviewees said that lower government stresses environmental issues that are of minor importance to them. This might be an important reason for a lack of congruence of governmental policy with business goals.

Table 7.10 demonstrates that the frequency of contacts to settle agreements on environmental issues in the chain between CEOs and different department managers is in general lower than once a year. Overall, the large firms indicate a higher frequency of contacts. The highest value is found for exchanges between the quality assurance departments, which can be explained by the fact that environmental issues may be agreed in line with food quality and safety issues (e.g. on the use of pesticides and chemicals). To follow up on this, it is interesting to explore the extent to which environmental cooperation is carried out.

Table 7.11 shows that the large compared to medium-sized firms pay less attention to environmental cooperation with buyers and suppliers, although no significant differences are found. One reason for this could be that they impose environmental requirements rather than search for environmental cooperation. Interestingly, no clear priority for sharing activities upstream or downstream the chain emerges from Table 7.11. Cooperation with buyers is only slightly more frequently carried out than cooperation with buyers. Overall, the figures show relatively low values. It might imply that environmental cooperation is typically taking place on an ad-hoc basis.

Table 7.12 shows that the small, medium-sized, and large firms reported significantly more frequently on the presence of a chain captain (e.g. large buyer or supplier) than the micro firms. This result is surprising to the extent that it was expected that large firms would act as a chain captain themselves rather than benefit from chain captain's assistance. However, a respondent from a large dairy processor (firm 4 in Table 6.1) indicates considering the firm to act as a chain captain as well: *We have substantively improved our energy efficiency over the past few years, which makes any further improvement hard to achieve. Yet, the government still focuses*

Table 7.10. Frequency of contacts to arrange agreements on environmental issues with buyers and/or suppliers (1 = never, 2 = yearly, 3 = twice a year, 4 = four times a year, 5 = monthly).

	Micro (N=196)	Small (N=177)	Medium-sized (N=81)	Large (N=26)
CEO/CEO	1.09 (0.45)	1.23 (0.71)	1.29 (0.72)	1.17 (0.48)
Purchasing/sales upstream	1.10 (0.48)[a,b,c]	1.33 (0.84)[c]	1.54 (1.01)[a]	1.88 (1.03)[b]
Sales/purchasing downstream	1.09 (0.50)[a,b]	1.25 (0.79)[e]	1.41 (0.81)[a]	1.62 (0.94)[b,d]
Marketing/marketing	1.06 (0.41)[e]	1.12 (0.51)	1.20 (0.50)	1.38 (0.82)[d]
R&D/R&D	1.09 (0.52)[a,e]	1.17 (0.66)[b]	1.31 (0.67)[c,e]	1.71 (1.00)[a,b,c]
Production/production	1.08 (0.52)[a,e]	1.21 (0.71)[b]	1.23 (0.62)[c,e]	1.33 (0.76)[a,b,c]
Logistics/logistics	1.07 (0.36)[a,c,e]	1.23 (0.75)[e]	1.35 (0.68)[a]	1.42 (0.78)[c]
Quality assurance/quality assurance	1.16 (0.68)[a,c,e]	1.41 (1.02)[d,e]	1.50 (0.91)[c]	1.92 (1.32)[a,d]

Significant differences are indicated with [a,b] $P < 0.01$; [c,d] $P < 0.05$; [e] $P < 0.10$

Table 7.11. Level of environmental cooperation with buyers and suppliers (1 = Not at all; 5 = Very much).

	Micro (N=196)	Small (N=177)	Medium-sized (N=81)	Large (N=26)
Buyers (upstream)	2.79 (1.26)	2.89 (1.16)	3.08 (1.10)	2.76 (1.17)
Suppliers (downstream)	2.74 (1.23)	2.80 (1.18)	2.88 (1.07)	2.72 (1.24)

Table 7.12. Presence of a supportive intermediary and participation in covenants (%).

	Micro (N=196)		Small (N=177)		Medium-sized (N=81)		Large (N=26)
Chain captain that provides help for environ. issues	19	—	28		35		31
Branch-org. that provides help for environ. issues	29	—	42	—	58		62
Participation in environmental covenants	46		52	—	84		81
One covenant only	43		39		52	—	27
Two or more	3	—	13	—	32	—	54

Differences between succeeding size categories are indicated by — (P < 0.05).

on energy reduction in our chain and, therefore, we are extending our attention towards suppliers to help them increase their energy efficiency as well.

The importance of branch-organizations to enhance information exchange in the sector is clearly indicated during the interviews. From the survey results, it appears that medium-sized and large compared to micro and small firms perceive more frequently support from a branch-organization (P<0.05). Although this result might partly depend on the availability of branch-organizations in the different sectors (see also Section 3.3.3, Box 3.4), firms not necessarily join a branch-organization, for instance, when it is perceived to lack effectiveness, such as indicated by the slaughterhouses interviewed (firm 7 and 8 in Table 6.1).

The level of participation in environmental covenants is investigated (see Table 7.12). A relatively large and significant difference is observed between the micro and small firms versus the medium-sized and large firms. A further distinction was made between firms that participated in one versus two or more covenants. Table 7.12 shows that the majority of large firms participates in two or more covenants: the Packaging Covenant is most frequently mentioned, followed by the Long-Term Agreement Energy.

7.3.3 Correlations and mean values

Spearman rank correlations and mean values are given for the micro and small firms in Table 7.13 and for the medium-sized and large firms in Table 7.14. As expected, the mean values show that environmental management capabilities (variable no. 1) are more developed in the medium-sized and large compared to micro and small firms (expressed on a 0-5 point scale). In line with the previous results, the micro and small firms pay very limited attention to the development of environmental management capabilities. Looking at the correlations, it can be noted that all coefficients are below 0.80, which indicates that no problems will be encountered in the regression analyses because of multi-collinearity (see Section 7.3.5). When comparing the results for the micro and small versus the medium-sized and large firms, the following three observations can be made.

First, the availability of a sufficient environmental budget (variable no. 4) shows the strongest significant correlation with the level of environmental capabilities in both the micro and small as well as medium-sized and large firms. It stresses the importance of financial resources in paying attention to the environment, not only in the smaller enterprises, but also in the larger firms.

Second, the contact frequency with lower government (variable no. 5) is strongly and significantly associated with the availability of environmental capabilities. Notably, the correlation is far stronger than that found between the governmental impact (variable no. 12) and the level of environmental management capabilities, which even appears non-significant in the medium-sized and large firms. This can be understood from the fact that all firms perceive the impact from government as being rather strong (see Table 7.7).

Third, the contact frequency with lower government (variable no. 5) seems to have different implications for the evaluation of the relationship with lower government in the micro and small firms versus medium-sized and large firms. In the micro and small firms, it is positively related to all other aspects of the relationship with government (variable no. 6 till 9), although not always significantly. Interestingly though, several of these aspects were negatively associated with the contact frequency in the medium-sized and large firms (variable no. 6, 7, and 9). Although not all of these correlations appear significant, it might indicate that the medium-sized and large firms perceive less benefit (e.g. help and support) from lower government than micro and small firms.

Table 7.13. Spearman rank correlations and mean values (± standard deviation) for the micro and small firms (N=195).

	Mean (S.D.)	1	2	3	4	5	6	7	8	9	10	11	12	13	14	15	16
1 Level of environ. capabilities	0.70 (0.86)	X															
2 Log(size)	1.07 (0.28)	**.40**	X														
3 Pollution level	2.60 (0.94)	**.35**	*.16*	X													
4 Environ. budget	2.02 (1.05)	**.49**	**.30**	**.33**	X												
5 Contact freq. government	1.86 (0.68)	**.47**	**.23**	**.29**	**.40**	X											
6 Quality of information	2.72 (0.93)	.10	.05	.10	.02	.17	X										
7 Importance as info source	2.85 (1.00)	.12	**.19**	.13	.11	.08	**.19**	X									
8 Equality and dialogue	2.45 (0.83)	**.21**	**.19**	.07	.13	**.32**	**.46**	**.20**	X								
9 Contrib. to business goals	2.66 (0.90)	.12	.13	.10	.06	.12	**.53**	.08	**.35**	X							
10 Contact freq. buyers/suppliers	1.25 (0.65)	**.27**	**.21**	.09	**.21**	**.21**	-.03	.02	.00	.06	X						
11 Cooperation buyers/suppliers	2.79 (1.14)	**.21**	*.18*	.09	**.20**	.12	**.19**	.02	**.22**	**.25**	**.33**	X					
12 Infl. government	3.63 (1.13)	**.23**	**.19**	.25	**.30**	.12	*-.16*	**.30**	-.04	-.04	.05	.07	X				
13 Infl. buyers	2.36 (1.21)	**.26**	.15	**.31**	**.25**	**.24**	-.06	*.18*	.01	.02	**.32**	**.36**	**.44**	X			
14 Infl. suppliers	1.90 (1.00)	.27	**.22**	**.33**	**.27**	*.16*	-.03	.12	.07	.11	**.32**	**.40**	**.33**	**.65**	X		
15 Infl. competitors	1.63 (0.77)	.32	.27	**.26**	**.31**	**.24**	-.03	.03	.05	.10	**.31**	**.30**	**.27**	**.60**	**.57**	X	
16 Infl. local inhabitants	2.65 (1.26)	**.21**	.13	**.42**	**.27**	**.27**	.07	.09	.09	.13	.09	.14	**.27**	**.33**	**.34**	**.27**	X
17 Infl. environ. organizations	1.92 (1.07)	*.18*	.06	**.23**	**.26**	**.23**	*.16*	.05	*.16*	.23	.05	**.21**	**.23**	**.29**	**.39**	**.31**	**.24**

Significant correlations are indicated in **Bold** (P < 0.01) and *Italic* (P < 0.05)

Table 7.14. Spearman rank correlations and mean values (± standard deviation) for the medium-sized and large firms (N=86).

	Mean (S.D.)	1	2	3	4	5	6	7	8	9	10	11	12	13	14	15	16
1 Level of environ. capabilities	2.26 (1.30)	X															
2 Log(size)	2.19 (0.31)	**.29**	X														
3 Pollution level	3.24 (0.92)	**.38**	.02	X													
4 Environ. budget	2.79 (0.83)	**.52**	**.28**	**.25**	X												
5 Contact freq. government	2.89 (0.90)	**.46**	.19	**.26**	.25	X											
6 Quality of information	2.99 (0.86)	.03	.11	-.17	**.32**	-.15	X										
7 Importance as info source	2.83 (0.78)	-.01	-.09	.05	.05	-.16	.14	X									
8 Equality and dialogue	2.94 (0.73)	.26	**.30**	.11	.27	.22	**.42**	.04	X								
9 Contrib. to business goals	2.81 (0.77)	.11	.04	.14	.03	-.19	.03	-.02	.15	X							
10 Contact freq. buyers/suppliers	1.50 (0.78)	**.34**	-.02	.17	.25	.14	-.01	.08	.08	.14	X						
11 Cooperation buyers/suppliers	2.94 (1.02)	**.31**	-.13	**.33**	.08	.14	-.02	.13	.11	.21	**.35**	X					
12 Infl. government	3.97 (0.79)	.08	-.19	**.37**	.00	.11	-.10	.11	.03	-.10	.07	.08	X				
13 Infl. buyers	2.71 (1.10)	.23	.03	.24	.19	.12	.08	.10	.17	.07	.27	**.44**	.01	X			
14 Infl. suppliers	2.02 (0.81)	.16	.09	.26	-.05	.02	-.01	.13	.18	.25	.21	.23	.04	**.54**	X		
15 Infl. competitors	1.84 (0.78)	.14	-.01	.18	.08	.19	-.04	.09	.08	.08	.14	.27	.25	**.38**	**.37**	X	
16 Infl. local inhabitants	3.06 (1.07)	**.37**	.02	**.49**	.25	.25	.02	-.12	.23	-.05	.12	.16	**.29**	**.28**	.05	.27	X
17 Infl. environ. organizations	2.08 (0.88)	**.34**	.03	**.33**	.19	.03	.06	.12	.22	.25	.18	.16	.21	.24	.23	**.43**	**.39**

Significant correlations are indicated in **Bold** (P < 0.01) and *Italic* (P < 0.05).

7.3.4 Regression analyses

Table 7.15 shows the results of the regression analysis for micro and small firms. It was decided to perform a binary logistic regression (see Section 5.6). A distinction is made between firms that implemented none (65% of the micro and small firms) versus one or more elements of the environmental management capabilities (35%). The estimates (see footnote of Table 7.18) indicate the direction of the relationship between the variables in terms of odds-ratios: values between 0 and 1 refer to a negative association, while values greater than 1 to a positive relationship. The bottom of the table provides information on the model fit and the model significance. All models were significant and the values of the Nagelkerke pseudo R^2 indicate sufficient predictive power.

The first model includes the control variables: it shows that company size, pollution level, and the availability of a sufficient environmental budget are all significant and positive predictors.

The second model concentrates on the aspects of the relationship with lower government. In line with the results of the correlation matrices, it appears that the frequency of contacts is the most important and only significant predictor, in addition to company size and pollution level. Remarkably, the frequency of contacts turns out to be a substitute, in strictly statistical terms, for the importance of the environmental budget. This might imply that the contacts with lower government help overcome problems of a limited financial budget, directly via environmental subsidies and/or indirectly via providing environmental information on how to deal with environmental regulations.

The third, fourth, and fifth models concentrate on chain actors, network actors, and societal groups, respectively. However, none of the variables measuring their impact appears to be significant. It may reflect, in general, a lack of environmental interests among stakeholders from the business network, other than government.

The final model includes all significant predictors found in the other models, which is, in fact, only the frequency of contact with lower government. It turns out to be a significant predictor. Interestingly, the environmental budget has become significant as well, although of minor importance (P<0.10).

Table 7.16 shows the results of the linear regression analysis for the medium-sized and large firms. Standardized regression coefficients are displayed (ranging between 0 and 1). The adjusted R^2 is sufficiently high for all models, which are all significant as well. A Kolmogorov-Smirnov test on the standardized residuals was carried out to statistically check their distribution. Both the test and a visual check indicate no significant deviations from a normal distribution, which is satisfactory. Furthermore, all VIF values are checked (not included in

Table 7.15. Binary logistic regression on the availability of environmental management capabilities in micro and small firms (N = 195)[1].

	1	2	3	4	5	Final
1. Control variables						
Log(size)	15.6**	13.0**	14.2**	16.7**	16.5**	15.5**
Pollution level	1.91**	18.1*	1.88**	1.95**	1.79**	1.71*
Environmental budget	1.66**	1.30	1.55*	1.69**	1.60*	1.45†
Dummy non-perishable food	0.57	0.71	0.70	0.57	0.54	0.76
2. Government						
Influence of government		1.29				
Frequency of contacts		2.91**				3.12**
Quality of information		0.86				
Importance as info source		0.93				
Equality and dialogue		1.02				
Contribution to own goals		1.13				
Dummy participation in one covenant only		0.23				
Dummy two or more covenants		0.01				
3. Chain actors						
Influence buyers			1.28			
Influence suppliers			0.72			
Frequency of contacts			1.66			
Cooperation buyers/suppliers			1.21			
Influence of competitors			1.66			
4. Network actors						
Dummy supportive branch-org.				0.83		
Dummy supportive chain captain				0.01		
5. Societal groups						
Influence local inhabitants					1.07	
Influence environmental org.					1.16	
Pseudo R^2	0.33	0.43	0.39	0.34	0.34	0.39
χ^2	54.4	73.3	65.6	55.2	55.5	66.3
P	0.00	0.00	0.00	0.00	0.00	0.00

** $P < 0.01$; * $P < 0.05$; † $P < 0.10$.

[1] Standardized estimates (exp. β) are displayed; a significant constant term is included in all models.

Table 7.16. Linear regression analysis on the availability of environmental management capabilities in medium-sized and large firms (N = 86)[1].

	1	2	3	4	5	Final
1. Control variables						
Log(size)	0.18†	0.11	0.17†	0.18*	0.18†	0.15†
Pollution level	0.27**	0.11	0.16	0.23*	0.17†	0.12
Environmental budget	0.39**	0.39**	0.40**	0.40**	0.35**	0.36**
Dummy non-perishable food	0.10	0.10	0.13	0.10	0.09	0.06
2. Government						
Influence of government		-0.01				
Frequency of contacts		0.30**				0.28**
Quality of information		-0.06				
Importance as info source		0.07				
Equality and dialogue		0.04				
Contribution to own goals		0.10				
Dummy participation in one covenant only			0.19			
Dummy two or more covenants		0.17				
3. Chain actors						
Influence buyers			-0.06			
Influence suppliers			0.13			
Frequency of contacts			0.11			
Cooperation buyers/suppliers			0.22*			0.21*
Influence of competitors			-0.02			
4. Network actors						
Dummy supportive branch-org.				0.08		
Dummy supportive chain captain				0.11		
5. Societal groups						
Influence local inhabitants					0.16	
Influence environmental org.					0.12	
R^2	0.37	0.48	0.44	0.39	0.41	0.48
Adj. R^2	0.34	0.39	0.37	0.35	0.37	0.44
F	11.9	5.6	6.5	8.5	9.1	12.2
df	85	73	76	79	79	79
P	0.00	0.00	0.00	0.00	0.00	0.00
ΔAdj. R^2	-	+0.05	+0.03	+0.01	+0.03	+0.10
ΔF	-	1.9†	1.8	1.6	2.7†	8.4**
K-S test on std. residuals	0.38	0.99	0.96	0.79	0.89	0.84

** $P < 0.01$; * $P < 0.05$; † $P < 0.10$.
[1] Standardized regression coefficients are displayed; a significant constant term is included in all models.

the table): they turned out to be below the upper threshold value of 10 so no multi-collinearity problems are encountered.

The focus of the different models is similar to the previous regression analysis for the micro and small firms. The first model shows that company size, pollution level, and the environmental budget are the significant control variables.

The second model shows that the contact frequency with lower government is the only significant predictor, in addition to environmental budget as a control variable. The fact that both these variables appear significant might indicate that contact with government, in contrast to the micro and small firms, does not help overcome limitations with respect to financial resources.

The third model evaluates the impact of chain actors. It appears that environmental cooperation with buyers and/or suppliers contributes significantly to the model. This is in line with the interviews, since it showed that several firms cooperate with buyers and suppliers to reduce environmental impacts, mainly related to obligations following from the Packaging Covenant.

The fourth and the fifth model evaluate the impact of network actors and societal groups, respectively. However, only the control variables turn out to be significant predictors in both models.

The final model comprises the significant predictors of the previous models. It shows that the contact frequency with lower government and environmental cooperation with buyers and/or suppliers are significant. Also company size and environmental budget appear to be significant, although the latter is of minor significance. The final model explains 44% of the variance in the adoption of environmental management capabilities.

7.4 Analysis of the 2005 survey data

This section presents the results of the second survey carried out. The data are gathered in 2005 from medium-sized and large Dutch food and beverage firms.

7.4.1 Environmental management capabilities and pollution

Table 7.17 shows the extent to which the medium-sized and large firms pay attention to environmental management capabilities. It shows that the large compared to the medium-sized firms have implemented more elements of the environmental management capabilities, although the differences are not always significant. The lowest values are found for the chain and cradle-to-cradle capabilities (no. 11 till 13). Notably though, the adoption has increased especially in the large firms compared to 2002. Section 7.5 elaborates in more detail on the

Table 7.17. Percentage of firms (%) implementing environmental management capabilities in the medium-sized and large firms in 2005 (N=100).

	Medium-sized (N=79)	Large (N=21)
Strategic capabilities		
1. Formulation of an environmental strategy	56	76 †
2. Formulation of an environmental action program	38	62 *
Operational capabilities		
3. Info. collection to check environmental emissions	71	86
4. Regular checks on environmental impact	49	91 **
5. Internal environmental information processing	57	81 *
6. Info. collection to evaluate environmental measures	52	76 *
7. Info. collection for internal environmental care	70	81
8. Environmental audit	50	71 †
9. Environmental database	43	57
10. Environmental training of employees	43	48
Chain and cradle-to-cradle capabilities		
11. Info. collection for chain-oriented environ. care	20	33
12. Info. collection to exchange with buyers/suppliers	9	33 **
13. Info. collection for environ. product (re)design	9	14

$** P < 0.01; * P < 0.05; † P < 0.10.$

longitudinal changes. In total, 12 firms were ISO14001 certified, including 2 from the meat, 5 from the vegetables and fruit, 1 from the dairy, 3 from the beverage sector and 1 processor from the category 'other'. Looking at the sector level in the first four cases, it equals 10%, 45%, 10%, and 38% of the firms, respectively. While it might indicate that the vegetables and fruit processors are more eager to get certified, the percentages should be interpreted carefully, since they are based on a limited number of firms only. Moreover, it might have been that certified firms were more willing to participate to show their environmental excellence: it appears that the availability of all items listed in Table 7.17 is significantly higher in the certified compared to the non-certified firms (P<0.01). However, two exceptions should be mentioned, namely the implementation of an environmental database (40% for certified versus 50% for non-certified firms) as well as environmental information collection for exchanges with buyers and/or suppliers (14% versus 17%). The first is surprising to the extent that a database is important to keep environmental records. It might be that the firms integrate environmental records in existing databases rather than establish a separate environmental database. Several interviewees indicated, for instance, that they aim to increase attention for the environment by including environmental issues in existing working routines for food quality and safety systems. The second exception, the fact that the ISO14001-certified firms do not carry out more frequently environmental information exchanges with other actors downwards and upwards in the chain, can be understood from the fact that the ISO14001 environmental management system is basically internally oriented. It is interesting, though, to note that the certified firms do pay significantly more attention to information collection for chain-oriented environmental care and environmentally responsible product (re)design. These two activities can be interrelated, since changes in products may have implications at the chain level.

Figure 7.2 shows the extent to which firms are dealing with different environmental issues. It demonstrates, as in 2002, that noise, water pollution, smell, and waste production are the most important topics. The level of environmental pollution is therefore calculated as the mean value of these four issues.

7.4.2 Firm and business network characteristics

Table 7.18 shows the mean values on firm strategy, enabling environmental capabilities as well as environmental image and environmental marketing. It shows that the large compared to medium-sized firms can be more clearly associated with a prospector strategy, since they attach significantly higher managerial importance to being the first to introduce new products. The third proposition suggests that a proactive attitude to being the first to launch new products implies that the firm is also more eager to pay attention to the environment. Looking at the empirical data, this association is found with respect to environmentally responsible product (re)design, see Figure 7.3.

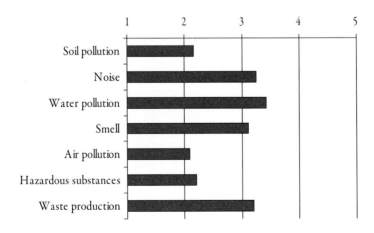

Figure 7.2. Environmental pollution impacts (1=Not important; 5=Very important).

Table 7.18. Firm strategy, enabling environmental capabilities, image and marketing (5-point Likert scales: higher values represent a more positive assessment; N=100).

	Medium-sized (N=79)	Large (N=21)
Prospector firm		
Our firm wants to be first to introduce new products	3.43 (1.24)	4.05 (1.15) *
Customers constantly ask for new products	3.40 (0.96)	3.60 (0.94)
Top-management commitment		
The board is actively involved in environ. management issues	3.77 (0.89)	3.80 (1.15)
Influence of environmental manager		
On strategic environmental issues	3.57 (0.96)	4.10 (0.68) *
On operational environmental issues	3.95 (0.68)	4.24 (0.83) †
Environmental communication		
Organizational culture that stimulates sharing of environ. ideas	3.44 (0.83)	3.90 (0.89) *
Environ. issues can be adequately comm. to higher management	3.85 (0.89)	4.10 (0.94)
Environmental information is shared among employees	3.35 (1.07)	3.57 (1.03)
Involvement of different departments		
Different departments are involved in environ. decision-making	3.16 (1.16)	3.70 (1.03) †
Environmental problems are collectively dealt with	3.95 (0.95)	4.38 (0.67) †
Integration of environ. issues in other management systems		
Quality Management	3.73 (1.23)	4.15 (1.18) †
Human Resource Management (HRM)	3.41 (1.16)	3.65 (1.42)
Image and marketing[1]		
Environmentally friendly image	2.57 (1.12)	3.42 (1.12) **
Environmental marketing opportunities	1.70 (0.92)	2.35 (1.09) **

** P < 0.01; * P < 0.05; † P < 0.10.
[1] The total number of observations is relatively low: image (N = 86) and marketing (N = 85).

Figure 7.3 illustrates that firms with a stronger prospector orientation pay relatively more attention to their product (re)design from an environmental point of view. This can be understood from their wish to differentiate from competitors. Examples of environmental issues that might be relevant in this context are re-usable packaging and environmentally friendly produced raw materials and manufacturing processes. No clear relationship is found between a prospector strategy and other elements of the environmental management capabilities.

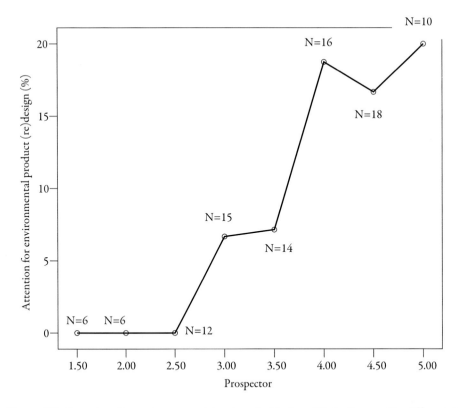

Figure 7.3. Attention for environmental product (re)design (%) in firms with different levels of prospector orientations.

Looking at the other variables in Table 7.18, it can be concluded that the environmental commitment of top management is rather positively evaluated. It suggests sufficiently high environmental awareness at the strategic level. Furthermore, it appears that the environmental manager has a relatively strong perceived impact on strategic and operational environmental activities, even significantly stronger in the large compared to the medium-sized firms. Rather high mean values are also found for the variables measuring internal environmental communication and the involvement of different departments.

Table 7.18 shows that the large compared to medium-sized firms turn out to integrate environmental issues more frequently in food quality and safety, as well as human resource management. They also indicate experiencing more benefits from their attention for the environment in terms of environmentally friendly image enhancement and marketing opportunities (P<0.01). This might be related to the fact that they attract more stakeholder interests, which have to be taken care of.

Table 7.19 shows that the large compared to the medium-sized firms perceive stronger impacts from their business network, although the differences are not significant for all stakeholders. The highest level of impact is measured for the national and lower government, while the EU appears to have a strong influence on the large firms in particular. This might be related to the fact that they are, in general, well informed about changes in EU regulations, which feeds the awareness of proposed or future environmental changes in European regulations. Looking at the other stakeholders, it turns out that branch-organizations have a high level of impact. This can be understood from their intermediary position between government and the firms, as well as their role in environmental information exchange between firms. In line with the interviews, Table 7.19 demonstrates that the perceived impact from the chain and network stakeholders is relatively low. Looking at the societal groups, local inhabitants appear to have a stronger impact than environmental organizations. This was also indicated by the interviewees (see Section 6.2.3). Furthermore, employees have a stronger influence in the large compared to the medium-sized firms, which might be due to an organizational culture that stimulates sharing environmental ideas and more frequent involvement of different departments in environmental issues. Longitudinal changes in the business network impact are discussed in Section 7.5.

Table 7.19. Business network impacts on environmental management (N=100).

	Medium-sized (N=79)	Large (N=21)
Government		
EU	3.16 (0.95)	3.68 (0.75) †
National government	3.51 (0.83)	4.11 (0.66) **
Lower government	3.80 (0.95)	3.95 (0.61)
Chain and network		
Branch-organization(s)	3.21 (0.96)	3.55 (0.83)
Buyers	2.80 (1.11)	3.25 (0.85) †
Suppliers	2.70 (1.07)	2.75 (0.91)
Consumers	2.56 (1.11)	2.86 (0.96)
Banks/insurers	2.57 (1.15)	2.70 (0.87)
Competitors	2.11 (0.92)	2.60 (1.05) †
Societal groups		
Local inhabitants	3.29 (1.12)	3.50 (0.89)
Environmental organizations	2.49 (1.00)	3.05 (0.87) *
Internal factors		
Employees	3.33 (0.86)	3.76 (0.77) †

** $P < 0.01$; * $P < 0.05$; † $P < 0.10$.

7.4.3 Correlations and mean values

Table 7.20 presents the mean values and Spearman rank correlations. The grey area includes the different stakeholder impacts. The mean value on the adoption of environmental management capabilities (variable no. 1) is slightly higher in 2005 compared to 2002. The other mean values are in line with the previously discussed results. Overall, all correlations appear to be lower than 0.80, which implies that no problems will be encountered in the regression analysis because of multi-collinearity. The correlation matrix shows several interesting results.

First, increased impact of an environmental manager (variable no. 8) is significantly and positively correlated to internal environmental communication (variable no. 6) and the involvement of different departments in environmental issues (variable no. 7).

Second, there is a remarkably low correlation coefficient between a prospector strategy (variable no. 4) and the adoption of environmental management capabilities. This might be due to the fact that prospector firms are particularly interested in environmentally friendly product (re)design (see Figure 7.3). A significant and strong correlation between a prospector strategy and the environmental commitment of top-management (variable no. 5) underpins the fact that prospector firms are dedicated to paying attention to the environment. The significant correlation between a prospector strategy and environmental marketing opportunities (variable no. 27) can be related to the prospector firms aiming to differentiate their products.

Third, it appears that an environmentally friendly image (variable no. 26) is significantly and strongly correlated to the availability of environmental management capabilities. This implies that an improved environmental reputation can be an important driver for paying attention to the environment. The fact that environmental marketing opportunities (variable no. 27) show a weaker correlation with the adoption of environmental management capabilities could be due to the limited environmental interests of buyers, as was indicated during the interviews (see Section 6.2.2).

Fourth, although a low average level of impact from banks and insurance firms (variable no. 24) is measured (see Table 7.19), their impact appears to be significantly correlated to the availability of environmental management capabilities. Interestingly, the impact of banks and insurance firms is significantly correlated with the perceived influence of competitors, branch-organizations, the EU, consumers, and local residents (variable no. 16, 17, 18, 22, and 23, respectively). It might mean that the environmental interests of bank and insurance firms increases with the attention paid to the environment in the business network as a whole.

Fifth, a remarkably negative relationship between the perceived impact of suppliers (variable no. 14) and the availability of environmental management capabilities exists (although not significant). This might be related to the fact that suppliers can hinder the supply of environmentally friendly raw materials. Furthermore, it might be difficult to make agreements

Table 7.20. Spearman rank correlations and mean values (± standard deviations) in medium-sized and large firms (N = 90).

	Mean (S.D.)	1	2	3	4	5	6	7	8	9	10	11	12	13	14	15	16	17	18	19	20	21	22	23	24	25	26
1 Environ. manag. cap.	2.29 (1.29)	X																									
2 Log(size)	2.17 (0.32)	**.47**	X																								
3 Pollution level	3.29 (0.83)	**.36**	**.28**	X																							
4 Prospector	3.53 (1.00)	.02	.13	.09	X																						
5 Top-management comm.	3.94 (1.02)	**.37**	*.23*	.08	**.43**	X																					
6 Environ. communication	3.59 (0.76)	**.45**	**.28**	**.29**	**.28**	**.39**	X																				
7 Involvement diff. depart.	3.61 (0.88)	**.54**	**.35**	**.35**	.13	.18	**.57**	X																			
8 Infl. environ. manager	3.79 (0.76)	**.48**	*.24*	.19	.11	**.30**	**.58**	**.60**	X																		
9 Integ. in quality man.	3.81 (1.21)	**.37**	*.21*	.18	.04	.12	**.43**	**.39**	*.22*	X																	
10 Integ. in HRM	3.46 (1.18)	*.23*	.16	.07	.06	.07	*.22*	*.21*	.18	**.42**	X																
11 Contrib. to business goals	2.84 (0.78)	*.21*	.02	**.30**	.11	-.03	.13	.12	.06	*.23*	*.23*	X															
12 Coop. buyers/suppliers	2.73 (0.99)	*.23*	.15	**.34**	*.21*	.16	**.57**	**.43**	*.23*	**.30**	*.25*	.19	X														
13 Number of covenants	1.67 (0.82)	**.44**	*.23*	.19	-.18	-.06	.15	**.28**	*.23*	*.23*	-.02	*.23*	.14	X													
14 Infl. suppliers	2.67 (0.98)	-.13	.05	*.22*	.11	**.32**	*.25*	.17	.08	.14	.19	*.21*	**.45**	-.01	X												
15 Infl. buyers	2.83 (1.06)	.10	.20	.18	*.27*	*.25*	.12	-.08	.04	.02	.05	*.22*	*.21*	.01	.20	X											
16 Infl. competitors	2.18 (0.97)	.11	.18	*.26*	.15	.07	.12	.06	-.02	.02	.04	**.29**	**.33**	-.01	*.24*	**.47**	X										
17 Inf. branch-org.	3.23 (0.95)	.17	.12	.20	.07	.09	.04	.07	.01	.03	*.21*	.06	**.29**	.14	*.26*	**.43**		X									
18 Infl. EU	3.21 (0.91)	.16	.20	*.22*	.20	*.25*	.10	.12	.15	-.09	-.12	.15	.17	.12	.17	.13	*.27*	**.46**	X								
19 Infl. national government	3.59 (0.82)	.19	.20	*.27*	*.26*	**.38**	.17	.12	.15	.15	.11	*.24*	*.21*	*.24*	.13	*.26*	*.23*	**.30**	**.60**	X							
20 Infl. local government	3.81 (0.91)	.18	-.02	.08	.09	-.04	.04	.01	.08	.16	.02	*.24*	-.15	.10	-.05	-.05	-.01	.02	.17	**.31**	X						
21 Infl. environ. org.	2.53 (0.91)	.13	.13	*.23*	.13	.10	.03	.11	.16	.11	.10	**.30**	.15	-.01	*.21*	.14	**.32**	*.22*	**.31**	**.29**	*.23*	X					
22 Infl. consumers	2.60 (1.09)	.00	.18	*.24*	**.37**	.20	.16	.00	-.02	-.05	.02	*.26*	*.25*	-.12	**.41**	**.50**	**.36**	*.27*	**.34**	.12	-.03	**.34**	X				
23 Infl. local residents	3.26 (1.08)	.15	.06	*.26*	*.26*	.11	.05	*.24*	.04	.08	.00	.07	.11	.09	.04	-.06	.12	*.24*	**.28**	**.34**	**.32**	**.32**	.16	X			
24 Infl. banks/insurance firms	2.58 (1.08)	**.29**	.05	*.23*	.17	.05	.12	.12	.11	-.02	.05	*.22*	.02	.11	.08	.09	**.30**	**.42**	**.30**	*.24*	.20	*.21*	**.28**	**.33**	X		
25 Infl. employees	3.42 (0.79)	.14	.08	**.35**	**.30**	*.27*	**.29**	**.28**	.18	.13	*.25*	.13	*.24*	-.06	*.26*	.08	*.23*	.20	.17	*.24*	.04	**.32**	*.21*	**.39**	**.29**	X	
26 Image	2.62 (1.09)	**.57**	**.32**	**.34**	.16	.15	**.46**	**.47**	**.35**	**.50**	.17	**.29**	**.36**	**.30**	.16	.14	.11	.14	.16	*.21*	.17	**.31**	*.23*	*.21*	.20	*.26*	X
27 Marketing	1.81 (0.92)	**.28**	*.22*	**.48**	**.30**	*.23*	**.43**	**.34**	.14	*.24*	.11	**.31**	**.53**	.11	**.32**	**.38**	**.36**	**.29**	**.35**	*.26*	.04	.20	**.38**	.11	.15	**.36**	**.43**

Significant correlations are indicated in **Bold** (P < 0.01) and *Italic* (P < 0.05).

with suppliers on environmental issues. Referring forward to the cluster analysis, however, it should be emphasized that this conclusion is rather premature. The cluster analysis in Section 7.4.5 demonstrates that the negative influence can be related to strong pressure from suppliers exerted on a specific group of firms that are (still) lagging behind in the development of environmental management capabilities.

7.4.4 Regression analyses

Table 7.21 includes the results of the regression analyses. The adjusted R^2 shows a sufficient predictive power of all models, which are all significant as well. The standardized residuals were visually checked: no problems because of non-normality are encountered. This is supported by the non-significant outcomes of the Komogorov-Smirnov test. The VIF values were assessed and they appeared to be lower than the upper threshold of 10. This indicates that no multi-collinearity problems exist in the models.

The first model in Table 7.21 includes the control variables, which are all significant.

The second model concentrates on firm strategy. In addition to the control variables, the influence of the environmental manager is significant (P<0.05). An interaction term with company size did not turn out significant, which suggests that its importance is not dependent on company size (see footnote of Table 7.21). In line with the correlation matrix, no significant result is found for a prospector strategy.

The third model deals with enabling environmental capabilities. Environmental communication and the involvement of different departments contribute significantly to this model. It was also statistically tested whether the importance of environmental communication and the involvement of different departments would be more important in the larger firms using an interaction term with company size (see footnote of Table 7.24). No significant result is found though. No significant result is found for the integration of environmental issues in existing food quality and safety systems and/or human resource management either. Beforehand it was argued that overly tight integration could hinder managerial flexibility in the different management areas (see Section 5.4.3). Since this would suggest a certain optimum in the level of integration, an inverse U-shaped distribution of the level of integration was included (see footnote of Table 7.24). However, no significant result is found.

The fourth model focuses on the influence of government. Participation in environmental covenants turns out significant. To check the most important environmental covenants, they are included as dummy variables as well. The Long-Term Agreement Energy (β=0.24; P<0.05), Energy Efficiency Benchmarking Covenant (β=0.16; P<0.10), and the Integral Environmental Task (β=0.21; P<0.05) contributed significantly to the model (see also Section 3.3.3, Box 3.4 for background on the covenants). Remarkably, the Packaging Covenant did not appear

Table 7.21. Regression analyses on environmental management capabilities (N = 90)[1].

	1	2	3	4	5	6	Final
1. Control variables							
Log(size)	0.34**	0.27**	0.20*	0.27**	0.32**	0.34**	0.18*
Pollution level	0.19*	0.16†	0.08	0.14	0.13	0.16	0.09
Dummy non-perishable food	0.24*	0.19*	0.22*	0.27**	0.18†	0.23*	0.17*
2. Firm strategy							
Prospector		-0.07					
Top-management commitment		0.13					
Infl. environmental manager[2]		0.26*					0.06
3. Enabling capabilities							
Environmental communication[2]			0.19†				0.18†
Involvement departments[2]			0.24*				0.18†
Quality management integration[3]			0.13				
HRM integration[3]			0.01				
Infl. employees			-0.00				
4. Government							
Influence EU				-0.03			
National government				-0.02			
Local government				0.08			
Contribution to own goals				0.06			
Number of covenants				0.34**			0.22*
5. Chain and network actors							
Infl. buyers					0.06		
Infl. suppliers					-0.23*		-0.24**
Cooperation buyers/suppliers					0.35**		0.09
Infl. of consumers					-0.15		
Infl. banks/insurance firms					0.29**		0.19*
Infl. competitors					-0.12		
Infl. branch-organizations					0.06		
6. Societal groups							
Infl. local inhabitants						0.09	
Infl. environmental organizations						-0.04	

Table 7.21. Continued.

	1	2	3	4	5	6	Final
R^2	0.31	0.42	0.48	0.44	0.47	0.32	0.60
Adj. R^2	0.29	0.38	0.43	0.38	0.41	0.28	0.55
F	12.9	9.9	9.3	7.8	7.1	7.8	11.8
df	86	83	81	81	79	84	79
P	0.00	0.00	0.00	0.00	0.00	0.00	0.00
ΔAdj. R^2	-	+0.09	+0.14	+0.09	+0.12	-0.01	+0.26
ΔF	- (-)	5.1**	5.2**	3.6**	3.5**	0.4	8.1**
K-S test on std. residuals	0.41	0.78	0.81	0.70	0.97	0.39	0.53

** $P < 0.01$; * $P < 0.05$; † $P < 0.10$.
[1] Standardized regression coefficients are displayed; a significant constant term is included in all models.
[2] An additionally included interaction term with Log(size) did not appear significant.
[3] An additionally included inverse U-shaped relationship did not appear significant.

significant, which might be related to the fact that 90% of the firms participated in it, which, in statistical terms, reduces variance.

The fifth model shows the influence of chain and network actors. Cooperation with buyers and/or suppliers and the impact of banks and insurance firms are significant variables. In line with the correlation matrix, it indicates that banks and insurance firms can have an important influence on firms to develop environmental management capabilities. Furthermore, a significant negative impact is found for suppliers, which is in concordance with the correlation results. It will be discussed in greater detail in Section 7.4.5.

The sixth model concentrates on societal groups: it shows no significant results for the impact of local inhabitants and environmental organizations on the availability of environmental management capabilities.

The final model includes all significant predictors of the previous models. It shows that environmental communication and the involvement of different departments in environmental issues are significant. The participation in covenants is a significant model parameter. The positive influence of banks and insurance firms as well as the negative impact of suppliers remain significant.

7.4.5 Cluster analysis

The significant determinants of the regression models in Table 7.21 were used as clustering variables (except for the control variables) in a hierarchical cluster analysis. Table 7.22 presents the final cluster centers. It shows an interesting breakdown of the sample into four groups of firms that can be characterized based on their level of environmental pollution and the development of environmental management capabilities.

The first cluster includes mainly medium-sized firms with a low urge to clean up (N = 20). This is concluded from the fact that they have the lowest level of pollution (variable no. 2), and (consequently) they also perceive a low level of business network impacts, including chain and network actors and societal groups in particular (variable no. 19 till 27). From this situation it follows logically that they give minor managerial priority to the development of environmental management capabilities (variable no. 3), other than required to act in conformance with environmental regulations.

The second (N = 20) and the third (N = 22) cluster consist of firms that have a medium level of environmental pollution. These two groups differ significantly from each other with respect to the level of adopted environmental management capabilities (see Table 7.22). The group with the lowest compared to the highest level of environmental management capabilities perceived a remarkably stronger impact from several chain actors, such as suppliers, buyers, and consumers (variable no. 14, 20, and 23). The environmental interest of these stakeholders downstream and upstream in the chain might be triggered by the lagging environmental management capabilities of the firms in the second cluster. Interestingly, the results indicate that the Board of Directors is committed to taking care of the environment (variable no. 7). The lagging performance is then perhaps due to difficulties in translating the strategic environmental ambitions into practice. A reason for this could be the limited influence of the environmental manager compared to the firms in the third cluster (variable no. 8).

The fourth cluster comprises firms (N = 17) with the highest level of pollution. This contributes to the urge to clean up. Their size and the environmental pressure might also be the reason for strong business network impacts. It is interesting to note that they pay more attention to a prospector strategy compared to the other cluster groups, although the differences are not significant (variable no. 6). The firms in the fourth cluster might try to proactively satisfy the green wishes of their customers by environmentally friendly product (re)design (see also Figure 7.9). The perceived benefit of this strategy is reflected in significantly greater levels of perceived environmental image enhancement (variable no. 4) and marketing opportunities (variable no. 5) compared to the other cluster groups.

Referring back to the correlation matrix and regression analysis (Table 7.20 and 7.21), the negative relationship between perceived impact from suppliers and the level of environmental management capabilities can be associated with the second cluster group. The firms in the

Table 7.22. Final cluster centers (N = 90)[1].

	Cluster 1 (N=20)	Cluster 2 (N=31)	Cluster 3 (N=22)	Cluster 4 (N=17)
Pollution level	Low	Medium	Medium	High
Environ. capabilities	Weak	Weak	Strong	Strong
Control variables				
1. Log(size)	2.00[a,c] —	2.17	2.21[a]	2.32[c]
2. Pollution level[1]	2.91[c]	3.26[b]	3.18 —	3.94[b,c]
Performance				
3. Level of environ. capabilities	1.37[a,b]	1.77[c] —	3.01[b]	3.37[a,c]
4. Image enhancement	1.88[a,c] —	2.50[b]	2.57[a] —	3.79[b,c]
5. Marketing opportunities	1.47[c]	1.78[b]	1.48 —	2.71[b,c]
Firm strategy				
6. Prospector firm	3.23	3.60	3.55	3.76
7. Top-management comm.	3.10[c] —	3.77[b]	3.59 —	4.71[b,c]
8. Infl. environmental manager	3.25[a,c] —	3.66[b]	4.05[a]	4.35[b,c]
Enabling capabilities				
9. Environ. communication[2]	2.87[a,c] —	3.51[b]	3.82[a] —	4.27[b,c]
10. Involvement diff. departments[2]	2.75[a,c] —	3.55[b]	3.80[a] —	4.50[b,c]
11. Integr. in quality management	3.25[c]	3.71[b]	3.91 —	4.53[b,c]
12. Integr. in HRM	3.05	3.55	3.36	3.88
13. Infl. employees	2.90[c] —	3.61	3.41	3.71[c]
Government				
14. Infl. EU	2.95[c]	3.23[b]	3.00 —	3.76[b,c]
15. Infl. national government	3.25[c]	3.55[b]	3.55 —	4.12[b,c]
16. Infl. local government	3.75	3.55[b]	3.86	4.29[b]
17. Contribution to business goals	2.55 —	2.94[b]	2.68 —	3.24[b]
18. Number of covenants[2]	1.20[a,c]	1.42[b] —	1.91[a]	2.35[b,c]
Chain actors				
19. Infl. suppliers[2]	2.00[c] —	3.45 —	1.64 —	3.35[c]
20. Infl. buyers	2.95	2.97 —	2.36 —	3.06
21. Cooperation buyers/suppliers	2.00[a,c] —	2.89[b]	2.59[a] —	3.50[b,c]
22. Infl. competitors	1.95[c]	2.23	1.91 —	2.71[c]
23. Infl. consumers	2.45[c]	2.90 —	1.86 —	3.18[c]
Network actors				
24. Infl. branch-organizations	3.05[c]	3.13[b]	3.14 —	3.76[b,c]
25. Infl. banks/insurance firms[2]	1.80[a,c] —	2.45[b]	2.82[a]	3.41[b,c]
Societal groups				
26. Infl. local inhabitants	2.60[a,c] —	3.35	3.50[a]	3.53[c]
27. Infl. environmental organizations	2.15	2.68	2.45	2.82

[1] Significant differences between categories are indicated with — and corresponding letters (P < 0.05).
[2] Clustering variables.

second cluster perceive great pressure from suppliers to clean up, while their environmental management performance is lagging. Table 7.23 shows the distribution of the different food and beverage sectors over the four cluster groups. It appears that the first cluster group with a low environmental impact consists of many medium-sized meat processors and bakery factories. Furthermore, the groups with a medium level of environmental impact have an equal percentage of meat processors, but the group with weak environmental management capabilities (cluster 2) is dominated by bakery factories (39%). Lastly, the group of firms with a high level of environmental impact (cluster 4) comprises firms from different sectors, including meat, bakery, vegetables and fruit, and dairy. Taking a closer look at the bakery sector (N = 26), it appears that 50% of the firms belongs to cluster 2. Furthermore, 29% of the bakery factories is included in cluster 1, 8% in cluster 3, and 13% in cluster 4.

Table 7.23. Distribution of sectors over the cluster groups.

	Cluster 1 (N=20)	Cluster 2 (N=31)	Cluster 3 (N=22)	Cluster 4 (N=17)
Pollution level	Low	Medium	Medium	High
Environ. capabilities	Weak	Weak	Strong	Strong
Meat	25%	23%	23%	12%
Bakery	35%	39%	9%	18%
Vegetables and fruit	10%	10%	5%	18%
Animal feed	-	3%	9%	-
Beverages	10%	7%	5%	6%
Dairy	5%	10%	14%	12%
Grain mill products	-	3%	9%	12%
Other	15%	5%	26%	22%
Total	100%	100%	100%	100%

7.5 Longitudinal analysis

This section presents the results of the longitudinal analysis of the medium-sized and large firms that participated in 2002 and/or 2005.

7.5.1 Baseline statistics

Figure 7.4 shows small increases in the perceived importance of different environmental issues from 2002 to 2005. Significant differences are found for soil pollution ($P<0.05$), emissions

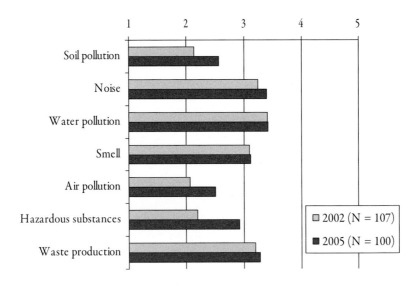

Figure 7.4. Perceived importance of pollution impacts in 2002 and 2005 (5-point Likert scale: 1 = Not important at all; 5 = Very important).

to air (P<0.01), and the release of hazardous substances (P<0.01). This might reflect an increased pressure on the environment, although the differences could also be due to increased environmental awareness. Several interviewees indicated that managerial priority for taking care of the environment has increased in recent years due to legal environmental demands and environmental initiatives carried out by branch-organizations.

Table 7.24 shows the longitudinal changes in the adoption of environmental management capabilities. In particular, several elements of the operational capabilities, such as information collection to evaluate environmental measures (variable no. 6) and for internal environmental care (variable no. 7) are more frequently implemented in 2005 compared to 2002. This might be related to the increased significance of environmental issues needing to be dealt with. Table 7.24 also indicates some rather remarkable decreases in elements of environmental management capabilities particularly in the medium-sized firms, which implemented significantly less frequent environmental audits in 2005 compared to 2002 (variable no. 8). Furthermore, it can be seen that the attention for chain and cradle-to-cradle capabilities increased in the large firms. While the differences are not significant, environmental information exchanges with buyers and/or suppliers and environmentally friendly product (re)design were carried out more often in 2005 than in 2002.

Table 7.25 shows that the average company size of the samples and the pollution level does not differ significantly between 2002 and 2005. In line with the previous findings, an increase in the adoption of environmental management capabilities is measured, although the differences

Table 7.24. The availability of environmental management capabilities in the medium-sized and large firms in 2002 and 2005.

	Medium-sized		Large	
	2002 **(N=81)**	**2005** **(N=79)**	**2002** **(N=26)**	**2005** **(N=21)**
Strategic capabilities				
1. Formulation of an environmental strategy	54	56	69	76
2. Formulation of an environmental action program	45	38	65	62
Operational capabilities				
3. Info. collection to check environmental emissions	67	71 †	77	86
4. Regular checks on environmental impact	58	49	69	91 *
5. Internal environmental information processing	49	57	69	81
6. Info. collection to evaluate environmental measures	42	52 †	62	76
7. Info. collection for internal environmental care	59	70 *	59	81 *
8. Environmental audit	61	50 *	58	71
9. Environmental database	41	43	58	57
10. Environmental training of employees	33	43 †	46	48
Chain and cradle-to-cradle capabilities				
11. Info. collection for chain-oriented environ. care	19	20	27	33
12. Info. collection to exchange with buyers/suppliers	11	9	23	33
13. Info. collection for environ. product (re)design	9	9	8	14

* $P < 0.05$; † $P < 0.10$.

Table 7.25. Firm characteristics and business network impacts.

	Medium-sized		Large	
	2002 (N=81)	2005 (N=79)	2002 (N=26)	2005 (N=21)
Control variables				
Log(size)	2.04 (0.20)	2.03 (0.19)	2.61 (0.17)	2.69 (0.19)
Pollution level	3.23 (0.83)	3.20 (0.81)	3.25 (1.02)	3.71 (0.88)
Performance				
Environ. management cap.	2.07 (1.31)	2.18 (1.32)	2.69 (1.39)	3.06 (1.14)
Government				
Infl. (lower) government[1]	4.03 (0.75)	3.80 (0.95)	3.69 (0.93)	3.95 (0.61)
Contribution to business goals	2.82 (0.77)	2.79 (0.76)	2.84 (0.80)	2.90 (0.94)
Chain actors				
Infl. buyers	2.66 (1.10)	2.80 (1.11)	2.96 (0.96)	3.25 (0.85)
Infl. suppliers	1.96 (0.72)	2.70 (1.07) **	2.29 (0.96)	2.75 (0.91) †
Environ. cooperation buyers	3.08 (1.10)	2.50 (1.03)	2.76 (1.17)	2.80 (1.15)
Environ. cooperation suppliers	2.88 (1.07)	2.92 (1.06)	2.72 (1.24)	3.33 (1.20)
Inf. competitors	1.88 (0.85)	2.11 (0.92)	1.79 (0.51)	2.60 (1.05) **
Societal groups				
Infl. local inhabitants	3.08 (1.12)	3.29 (1.12)	3.13 (0.95)	3.50 (0.89)
Infl. environ. organizations	2.16 (1.03)	2.49 (1.00) **	2.17 (0.87)	3.05 (0.87) **

** P < 0.01; † P < 0.10.
[1] The impact of lower government was taken in 2005.

are not significant. Overall, the results shows higher levels of business network impacts in 2005 compared to 2002. An exception is found for the governmental impact in the medium-sized firms (see Table 7.25): it declined, although not significantly. It is also interesting to note that the contribution of governmental environmental policy goals to business goals hardly changed from 2002 to 2005. Looking at the chain actors, a significant increase in influence is found for the suppliers. This might reflect the growing interest of these stakeholders in environmental issues from a chain perspective, which is underpinned by increased levels of environmental cooperation with suppliers in 2005 compared to 2002 in both the medium-sized and large (although the differences are not significant). It is interesting to note that the impact of competitors increased, as well, although only significantly in the large firms. Furthermore, a significant increase in impact is measured for environmental organizations.

Table 7.26 demonstrates that fewer firms perceive environmental support from a branch-organization in 2005 compared to 2002, which is significant for the medium-sized firms. At the same time, though, the number of firms that participated in public-private environmental voluntary agreements is significantly higher in 2005 compared to 2002.

7.5.2 Correlations and mean values

Table 7.27 shows the correlations coefficients and the mean values of the research variables for the total sample of medium-sized and large firms. In line with the previous results, firms perceive the strongest impact from government (variable no. 4) and local inhabitants (variable no. 10). The strongest correlations with the adoption of environmental management capabilities (variable no. 1) are found for company size (variable no. 2) and pollution level (variable no. 3). Also environmental cooperation with buyers and/or suppliers (variable no. 8) and perceived influences from local inhabitants (variable no.10) are significantly correlated to the availability of environmental capabilities.

All correlations are below 0.80, which implies that no multi-collinearity problems are to be expected in the regression analysis.

7.5.3 Regression analysis

Table 7.28 shows the results of the linear regression analysis. All models are significant and the adjusted R^2 indicates a sufficient predictive power. The distribution of the standardized residuals is visually evaluated and checked by a Kolmogorov-Smirnov (K-S) test, which both

Table 7.26. Presence of a supportive branch-organization and participation in covenants (%).

	Medium-sized		Large	
	2002 (N=81)	2005 (N=79)	2002 (N=26)	2005 (N=21)
Branch-org. that provides help for environ. issues[1]	58	46 *	62	55
Participation in environmental covenants				
One covenant only	52	44	27	24
Two or more covenants	32	52 **	54	76 *

** P<0.01; * P<0.05.

[1] In 2005, the influence of branch-organizations was used as a proxy for perceived environmental support: 1-2 = no supportive branch-organization; 3-5 = supportive branch-organization.

Table 7.27. Spearman rank correlations and mean values (± standard deviation) (N=187).

	Mean (S.D.)	I	2	3	4	5	6	7	8	9	10
I. Environ. manag. cap.	2.30 (1.31)	X									
2. Log(size)	2.18 (0.32)	.36	X								
3. Pollution level	3.27 (0.87)	.36	.13	X							
4. Infl. (lower) government	3.87 (0.85)	.20	-.08	.23	X						
5. Contribution to own goals	2.81 (0.79)	.15	.01	.23	.08	X					
6. Infl. buyers	2.78 (1.08)	.16	.12	.19	-.01	.10	X				
7. Infl. suppliers	2.38 (0.99)	.05	.08	.21	.01	.18	.35	X			
8. Coop. buyers/suppliers	2.85 (1.00)	.29	.05	.34	-.02	.16	.33	.33	X		
9. Infl. competitors	2.02 (0.90)	.16	.10	.20	.14	.16	.44	.37	.32	X	
10. Infl. local inhabitants	3.19 (1.08)	.28	.02	.37	.31	-.00	.10	.10	.14	.22	X
II. Infl. environ. org.	2.32 (0.95)	.24	.10	.23	.24	.25	.21	.30	.19	.41	.35

Significant correlations are indicated with **Bold** (P < 0.01) and *Italic* (P < 0.05).

indicate no significant deviations from a normal distribution. Multi-collinearity is checked by evaluating the VIF values. They were all below the upper threshold value of 10, which is satisfactory.

The first model includes the control variables. A dummy variable is used for the year of measurement. In contrast to the other variables, it does not contribute significantly to the model.

The second model concentrates on the governmental impact. The perceived influence from government turns out to contribute significantly to the model, although it is of minor importance (P<0.10). The participation in two or more environmental covenants is significant as well (P<0.01).

The third model focuses on the chain and network actors. Cooperation with buyers/suppliers appears to be a significant predictor. Furthermore, the dummy for the year of measurement is significant, although of minor importance (P<0.10). This result corresponds with the observed differences between the years in different stakeholder influences (see Table 7.25) and the presence of supportive branch-organizations (see Table 7.26).

The fourth model includes societal groups. It shows that the perceived influence of local inhabitants is a significant predictor.

Table 7.28. Linear regression analysis on environmental management capabilities in the medium-sized and large firms (N =187)[1].

	1	2	3	4	Final
1. Control variables					
Log(size)	0.29**	0.25**	0.28**	0.29**	0.25**
Pollution level	0.33**	0.24**	0.26**	0.25**	0.17**
Dummy non-perishable food	0.17**	0.18*	0.18*	0.15*	0.18*
Dummy year	0.06	0.01	0.12†	0.02	0.03
2. Government					
Influence (lower) government		0.12†			0.11
Contribution to own goals		0.03			
Dummy participation in one covenant only		0.02			
Dummy two or more covenants		0.27*			0.21**
3. Chain and network actors					
Influence buyers			0.03		
Influence suppliers			-0.11		
Cooperation buyers/suppliers			0.22**		0.16**
Competitors			-0.01		
Dummy supportive branch-organization			0.09		
4. Societal groups					
Influence local inhabitants				0.15*	0.09
Environmental organizations				0.08	
R^2	0.27	0.35	0.33	0.30	0.38
Adj. R^2	0.25	0.32	0.29	0.28	0.35
F	16.8	11.8	9.5	12.9	13.4
df	182	178	177	180	178
P	0.00	0.00	0.00	0.00	0.00
ΔAdj. R^2	-	+0.07	+0.04	+0.03	+0.10
ΔF	-	5.2**	2.9*	4.0**	7.5**
K-S test on std. residuals	0.32	0.89	0.70	0.61	1.00

** $P < 0.01$; * $P < 0.05$; † $P < 0.10$.
[1] Standardized regression coefficients; all models included a significant constant term.

The final model includes the significant predictors of the previous models. It appears that the control variables contribute significantly to the final model, except for the year of measurement. It appears that both the participation in two or more environmental covenants and the level of environmental cooperation with buyers/suppliers are significant predictors for the level of environmental management capabilities.

7.6 Concluding remarks

This chapter discussed the empirical findings acquired from 492 Dutch food and beverage firms in 2002 (including 106 medium-sized and large firms) and 100 medium-sized and large Dutch food and beverage firms in 2005. A longitudinal analysis provided insight into the changes in the adoption of environmental management capabilities and perceived business network impacts from 2002 to 2005.

Table 7.29 shows the confrontation of the propositions with the empirical data. The results of the micro and small firms are evaluated first. No evidence is found to support the first proposition (P1), which suggests a positive effect of environmental covenants on the adoption

Table 7.29. Empirical assessment of the propositions for the micro, small, medium-sized and large firms.

Focus of the proposition	Micro and small	Medium-sized and large	
	2002	2002	2005
(P1) Environmental voluntary agreements	0	0	+++
(P2) Contact intensity with lower government	+++	+++	a
(P3) Environmental support from an intermediary	0	0	a
(P4) Pressure from chain actors	+	+	+++
(P5) Prospector strategy and cradle-to-cradle capabilities	a	a	++
(P6) Top-management commitment	a	a	+
(P7) Perceived impact of an environmental manager	a	a	++
(P8) Horizontal and vertical communication	a	a	+++
(P9) Availability of (other) care systems	a	a	+

The following criteria are applied: +++ = significant according to the final regression model; ++ = significant according to a sub-model of the regression analysis or graph; + = significant according to the correlation analysis; 0 = no significant result found in the regression or correlation analysis.

of environmental management capabilities. The second proposition (P2) is confirmed, since the contact frequency with lower government is a significant predictor in the final regression models (see Table 7.15 and 7.16). In addition, the correlation coefficients indicate that contacts with lower governmental agencies can contribute to gaining valuable environmental information (see Table 7.13). No significant results are found to support the third proposition (P3): the presence of a supportive intermediary (e.g. branch-organization or chain captain) does not appear significant in the regression analysis. The fourth proposition (P4) focuses on the perceived chain pressures. Although the regression analysis indicates no significant results for the impact felt from chain actors (e.g. buyers and suppliers), the impact of chain actors is significantly correlated to the adoption of environmental management capabilities (see Table 7.13).

Looking at the medium-sized and large firms, some differences are encountered between 2002 and 2005 (see Table 7.29). Although no empirical evidence is found in 2002 to support the first proposition (P1), the regression analysis demonstrates a significant contribution of the participation in environmental covenants to the adoption of environmental management capabilities in 2005 (compare Table 7.16 and 7.21). The second proposition (P2) is confirmed, since the contact frequency with lower government appears to be a significant predictor for the attention paid to the environment (see Table 7.16). The third proposition (P3), which focuses on the role of intermediary organizations, is not confirmed by the survey data. In line with the fourth proposition (P4), suppliers appear a significant predictor for the adoption of environmental management capabilities in 2005 according to the regression analysis (see Table 7.21). Although their impact turns out to be negative according to the regression analysis, the cluster analysis demonstrates that this result is related to strong pressure felt from suppliers, while the adoption of environmental management capabilities is (still) lagging behind (see Table 7.22). Furthermore, the cooperation with buyers and suppliers turns out to be a significant predictor for the availability of environmental management capabilities in 2002 (see Table 7.16).

The fifth proposition (P5) refers to increased attention for cradle-to-cradle environmental capabilities in prospector firms. This assumption is confirmed by Figure 7.3, which demonstrates a positive relationship between a prospector strategy and environmental product (re)design. Furthermore, the cluster analysis reveals that the firms with the strongest developed environmental management capabilities are prospectors (although the differences with the other firms are not significant; see Table 7.22). It should be added, though, that no significant outcomes are found for a prospector strategy in the correlation matrix and regression analysis (see Table 7.20 and 7.21). This is probably due to the low number of 10 prospector firms (see also Figure 7.3). Limited empirical evidence was found for the sixth proposition (P6): although top-management commitment is significantly and positively correlated to the adoption of environmental management capabilities, it is not significant in the regression analysis. In line with the seventh proposition (P7), a sub-model of the regression analysis confirms the importance of an environmental manager (see Table 7.21).

Besides, the cluster analysis indicates that the environmental manager has a stronger impact in firms with more developed environmental management capabilities as well (see Table 7.22). Internal environmental communication and the involvement of different departments turn out to be significant predictors in the final regression model, which is in accordance with proposition eight (P8). Finally, limited empirical evidence is found for proposition nine (P9), which proposes a positive effect of existing care systems on the development of environmental management capabilities. The correlation matrix shows significant results for the integration of environmental issues in both quality and human resource management systems (see Table 7.20). These variables do not, however, appear significantly in the regression analysis.

8. Discussion and conclusions

This chapter discusses the empirical findings and draws conclusions. Section 8.1 elaborates on the adoption of environmental management capabilities in 2005 compared to 2002. Section 8.2 focuses on the business network and the changes in stakeholder influences between 2002 and 2005. Section 8.3 concentrates on firm strategy and Section 8.3 assesses the importance of enabling capabilities. Section 8.4 answers the central research question by evaluating the joint impact of the business network and firm characteristics on the adoption of environmental management capabilities. It focuses on the results of the cluster analysis, which resulted in four environmental profiles. Section 8.5 then evaluates the theoretical contributions of the present study and the methodological implications. Section 8.6 provides suggestions for further research. Section 8.7 discusses the managerial and policy implications.

8.1 Environmental management capabilities

The micro and small firms developed fewer environmental management capabilities than the medium-sized and large firms (see Table 7.5). This result is in line with previous studies that found minor interest small firms got environmental management (Hillary, 1999; Del Brío and Junquera, 2003; Hillary, 2004). This can be understood from the fact that they have, in general, a low environmental impact and, therefore, a low urge to pay attention to the environment. Moreover, micro and small firms are of a limited scale, which reduces the need to formally organize environmental activities. In the medium-sized and large firms, environmental management capabilities were more developed in 2005 compared to 2002 (see Table 7.24). In particular, the attention for operational aspects increased, which is reflected in the fact that more firms carried out regular checks on environmental impacts and information collection for internal environmental care. Control of own environmental performance is a prerequisite for being able to expand environmental attention beyond the firm (Lippmann, 1999). Hence, the development of operational environmental management capabilities can be interpreted as conditional for future improvement of chain-oriented and cradle-to-cradle environmental capabilities. Whilst not significant, the adoption of these capabilities already slightly increased in large firms, since environmental information exchange with buyers and/ or suppliers increased from 2002 to 2005. The attention for environmentally responsible product (re)design also grew especially in prospector firms. ISO 14001-certified firms (12% of the firms in 2005) appeared to pay significantly more attention to strategic and operational environmental management capabilities than the other firms. Interestingly they also turned out to pay significantly more attention to environmental activities at the chain level and to environmental product (re)design. So, ISO14001 certification, although predominantly internally oriented, seems to support product stewardship at the chain level, as well. A pilot study for the Dutch chemical industry supports this finding (VNCI, 2003).

8.2 Business network impacts

An increased impact of the business network is measured from 2002 to 2005, which is in accordance with a general observation of growing attention for the environment. The significance of banks and insurance firms in particular in the final regression model (see Table 7.21) is in line with recent debates on the role of financial institutions in stimulating industry to green their business[14]. There is clearly growing attention for environmental issues as part of the corporate social responsibility (CSR) strategy of banks and insurance firms, which are increasingly confronted with pressure from (inter)nationally organized societal groups who expose them for a lack of attention regarding social and environmental issues. For instance, the Friends of the Earth Netherlands recently showed that Dutch banks invest over twenty times more in projects that harm the climate, such as oil extraction and coal-fired power plants, than in projects for renewable energy (Milieudefensie, 2007). Box 8.1 includes the recommendations that were addressed.

A greater impact from chain actors and competitors is perceived in 2005 compared to 2002 (see Table 7.25). This might imply that the attention for environmental issues in the business networks has extended from government and environmental organizations to include stakeholders that are of primary interest for attaining commercial business goals. In line with the premises of the natural resource based view (natural-RBV) this implies that firms have to increasingly address environmental issues to fulfill the demands of their primary stakeholders. This growing number of stakeholders with environmental concerns will create a new level playing field. The fact that environmental cooperation with suppliers increased in the large firms in particular, although not significantly (see Table 7.25), can be interpreted as an important condition on which to base chain-oriented and cradle-to-cradle environmental capabilities.

Also the perceived impact of environmental organizations turned out to be higher in 2005 compared to 2002 (see Table 7.25). It might reflect the growing environmental concerns of society and also the increased success of environmental organizations in getting media attention for environmental concerns, such as the previously addressed initiative by the Friends of the Earth International (see Box 8.1). The present study confirms, however, that government is (still) the most dominant environmental stakeholder. This result is completely in line with previous studies (Braglia and Petroni, 2000; Madsen and Ulhøi, 2001). It can be explained by the fact that firms are legally obliged to comply with environmental regulations. However, the present study adds insight into the public-private relationship by showing the importance of public-private voluntary environmental agreements and the contact intensity with government. Table 8.1 assesses the propositions on the business network impact, which are discussed below.

---

[14] *NRC* June 1 2007.

Box 8.1. Attention for environmental issues in investment portfolios.

The following recommendations were made by the Dutch Chapter of Friends of the Earth International based on their investigation of attention for climate issues in the investment portfolios of banks.

To banks:
- Recognize, measure, and report your CO_2 emissions that result from loans and investments;
- Reduce investments in projects and firms that cause CO_2 emissions.

To governments:
- Establish a binding and uniform legal framework for banks to report on CO_2 emissions that result from their loans and investments;
- Establish a reporting point for socially unacceptable transactions where Non-Governmental Organizations (NGOs) and affected stakeholders can file complaints on what they see as socially unacceptable bank transactions;
- Promote European legislation to ensure responsible investment practices by all European banks, thereby providing a level playing field.

To consumers:
- Switch to a climate-conscious bank for your personal banking services;
- Ask others to move to a climate-conscious bank;
- Invest in sustainable energy funds and make use of "climate products" such as climate-friendly mortgages.

To civil society organizations:
- Start talking to your organizations' banks about integrating climate concerns in their loan and investment practices

---

Adapted from: Milieudefensie (2007).

P1 The participation in environmental voluntary agreements will be positively related to the adoption of environmental management capabilities.

Table 8.1 shows that no empirical evidence is found for the first proposition (P1) in 2002, while the interviews and the results of the survey in 2005 confirm the significance of participation in environmental voluntary agreements (or: covenants). The number of medium-sized and large firms involved in two or more covenants increased significantly from 2002 to 2005 (see Table 7.26). Several interviewees indicated that the participation in public-private environmental voluntary agreements is fruitful, since they get valuable feedback and it provides the opportunity to benchmark the environmental performance of the firm

Table 8.1. Confrontation of the propositions with the empirical findings on business network impacts.

Focus of the proposition	Micro and small	Medium-sized and large		
	2002	2002	2005	Interviews
(P1) Environmental voluntary agreements	0	0	+++	++
(P2) Contact intensity with lower government	+++	+++	a	+
(P3) Environmental support from an intermediary	0	0	a	+++
(P4) Pressure from chain actors	+	+	+++	+

+++ = strong confirmation; ++ = confirmation; + = limited empirical evidence; 0 = no significant result; a = not included in the analysis. For an explanation on the applied criteria, see Table 6.5 and 7.29 in Section 6.4 and 7.6, respectively.

against others (see Table 6.2). Interestingly, though, several respondents regard environmental agreements as a weak form of enforced regulation. They expect that, if they did not participate, lower government would impose the environmental targets included in the covenants anyway. Furthermore, some respondents indicated that because of collective reporting in some sectors, there is a chance of groups of firms that lag behind in environmental performance, unfairly benefiting from other firms showing environmental excellence.

P2 The contact intensity with lower government will be positively related to the adoption of environmental management capabilities.

Referring to the second proposition (P2), the correlation coefficients and regression models provide empirical evidence that the contact intensity with lower government is a significant predictor for the adoption of environmental management capabilities (see Table 7.13 till 7.16). Interestingly, contacts with government turn out to have different implications for the micro and small compared to the medium-sized and large firms. The correlation coefficients suggest that micro and small firms can benefit from governmental contacts by obtaining valuable information. By contrast, it is negatively correlated to the perceived quality of the provided environmental information by government in the case of medium-sized and large firms (although not significant, see Table 7.14). An explanation for this is found in the interviews, since a number of respondents stated that they seldom get new information from lower governmental officials but that they use them to discuss environmental permits and to try to adjust requirements to the specific situation of the firm (see Section 6.2.1). In this perspective, the contact intensity with lower government might reduce the chance of incongruence with the governmental environmental requirements and business priorities. Furthermore, it can be added that the interviewees of large firms felt that they acted as an important information source for lower governmental officials, especially concerning new environmental regulations at the European level.

P3 Environmental support from an intermediary, such as a branch-organization, will be positively related to the adoption of environmental management capabilities.

The third proposition (P3) evaluates the importance of environmental support from an intermediary. Although the surveys do not provide significant results to confirm this proposition, the interviewees clearly indicate that branch-organizations play a crucial role in enhancing environmental attention through facilitation of environmental information exchange, not only between government and firms, but also among firms in one sector (see Table 6.3). Several respondents confirmed, for instance, that their participation in environmental covenants strongly depends on the commitment of their branch-organizations. It should, however, be noted that the ability of branch-organizations to participate in environmental agreements will, of course, depend on the willingness of their member-firms to cooperate as well.

P4 Pressure from chain actors to pay attention to the environment will be positively related to the adoption of environmental management capabilities.

The fourth proposition (P4) concentrates on the perceived pressure from chain actors. Limited empirical evidence was found in 2002, while the regression analysis demonstrates that the impact felt from suppliers is a significant predictor for the development of environmental management capabilities in 2005 (see Table 7.21). However, the impact of the perceived pressure is negative in the regression model. This may indicate that suppliers, in general, have less power to influence the behavior of their buyers, for large buyers can, in general, choose other suppliers. In particular, this finding can further be explained by the cluster analysis showing that a large group of mainly bakery factories perceive strong pressure from their suppliers, whereas the development of environmental management capabilities is (still) lagging behind (see Table 7.22). These bakery factories are what Pavitt (1984) calls supplier-dominated firms, since they heavily rely on large powerful consortia that collectively buy ingredients. In line with the expectations of an interviewee of a medium-sized bakery factory, these consortia are increasingly paying attention to the environment to safeguard an environmentally friendly reputation further downstream in the chain (e.g. among retailers and consumers).

8.3 Firm strategy

Firm strategy encompasses both business and environmental strategy. The first is evaluated by assessing to what extent a firm is pursuing a prospector strategy, while the latter is considered on the environmental commitment of top management and the influence of an environmental manager on environmental issues. Table 8.2 includes the assessment of the propositions.

Table 8.2. Confrontation of the propositions related to corporate strategy with the empirical findings in 2005.

	Survey 2005	Interviews
(P5) Prospector strategy and cradle-to-cradle capabilities	++	++
(P6) Top-management commitment	+	+++
(P7) Perceived impact of an environmental manager	++	+++

+++ = strong confirmation; ++ = confirmation; + = limited empirical evidence. For an explanation on the applied criteria, see Table 6.5 and 7.29 in Section 6.4 and 7.6, respectively.

P5 A prospector strategy will be positively related to the adoption of environmental management capabilities, in particular cradle-to-cradle capabilities (e.g. environmentally friendly product (re)design).

The fifth proposition (P5) is confirmed by the survey results, since there is a clear positive relationship between the attention paid to environmental product (re)design and the implementation of a prospector strategy (see Figure 7.3). Several interviewees stated that the proactive business strategy stimulates environmental excellence (see Section 6.3.1). However, the regression analysis does not confirm a significant relationship between a prospector strategy and the adoption of environmental management capabilities, in general. This is in line with the fact that we found that prospectors pay more attention to cradle-to-cradle capabilities than other firms, but not to other environmental capabilities. Besides, a lack of statistical evidence might be related to the low number of prospectors firms (only 10 included in the sample).

P6 Strong top-management commitment to the environment will be positively related to the adoption of environmental management capabilities.

The sixth proposition (P6) concentrates top-management commitment. Although it is not confirmed by the regression analysis, a significant correlation is found between top-management commitment and the adoption of environmental capabilities (see Table 7.20). In addition, the interviews provide clear empirical evidence in favor of this proposition, since the respondents indicated that the commitment of the Board of Directors is essential for structurally enhancing environmental care. This is completely in line with previous studies (Roy *et al.*, 2001; Lee and Ball, 2003; Govindarajulu and Daily, 2004). Furthermore, several respondents emphasized the importance of a board member taking care of the stakeholder relationships related to environmental issues.

P7 A strongly perceived impact of the environmental manager on operational and especially strategic environmental issues will be positively related to the adoption of environmental management capabilities.

The seventh proposition (P7) is confirmed by a sub-model of the regression analysis in which the perceived impact of the environmental manager is a significant predictor for the availability of environmental management capabilities (see Table 7.21). In the interviews the environmental managers indicated that a significant part of their job was dedicated to safeguard the environmental awareness in the organization through regular contacts with the different departments and organizational levels (see Section 6.3.3). This is especially important for the implementation of environmental measures, which do not give distinct positive returns to the firm, such as the reduction of noise and smell.

8.4 Enabling environmental capabilities

Table 8.3 shows the confrontation of the propositions with the empirical findings regarding the enabling of environmental capabilities.

P8 Higher levels of horizontal (e.g. at the same organization level) and vertical (i.e. bottom-up and top-down) communication will be positively related to the adoption of environmental management capabilities.

The eights proposition (P8) focuses on the importance of internal communication. It is supported by the final regression model which suggests that the establishment of sufficient environmental communication channels is important to enhance employee environmental feedback (see Table 7.21). Interestingly, most interviewees indicated having more frequent information exchanges on environmental issues with production and logistics than with the marketing and sales departments (see Section 6.3.3). This is in line with a focus on the control of environmental impacts in the interviewed firms. The finding is also in accordance with the

Table 8.3. Confrontation of the propositions on enabling environmental capabilities with the empirical findings in 2005.

	Survey 2005	Interviews
(P8) Horizontal and vertical communication	+++	+++
(P9) Availability of (other) care systems	+	++

+++ = strong confirmation; ++ = confirmation; + = limited empirical evidence. For an explanation on the applied criteria, see Table 6.5 and 7.29 in Section 6.4 and 7.6, respectively.

limited attention for chain-oriented and cradle-to-cradle environmental capabilities, which require the involvement of the R&D, marketing, and the sales department.

P9 The availability of (other) care systems (e.g. on food quality and safety) will be positively related to the adoption of environmental management capabilities.

The ninth proposition (P9) is evaluated by the integration of environmental issues in existing care systems. Although a significant correlation is found with the adoption of environmental management capabilities (see Table 7.20), it does not turn out to be significant in the regression analysis. The majority of the interviewees explained, however, that they have integrated environmental and food quality issues by operational measures in particular, including joint audits, performance records and instruction guidelines or handbooks (see Section 6.3.4). Food quality and safety issues are, nevertheless, more important for them than environmental issues.

8.5 The way to environmental excellence

This book aimed to provide a deeper understanding of the factors that have an impact on the adoption of environmental management capabilities in firms in the Dutch food and beverage industry. The following research question is therefore put central. *What is the joint impact of the business network (government, buyers, suppliers, customers, etc.) and firm characteristics on the adoption of environmental management capabilities in Dutch food and beverage firms?* Although the previous section of this chapter has already discussed parts of this question, an answer with respect to the collective impact of the research variables can be found in the cluster analysis (see Table 7.22). The results demonstrate that four groups of firms can be distinguished based on their pollution level and the development of environmental management capabilities.

Following the environmental strategy typologies by Roome (1994), one cluster can be referred to as compliance-oriented (see Box 8.2). It is characterized by a low level of environmental pollution and, consequently, minor pressure from the business network to pay attention to the environment. In fact, the most important stakeholder appears to be lower government to safeguard the legal *license-to-produce*. Under the circumstances of a low environmental pollution level, it can be understood that the compliance-oriented firms have only weakly developed environmental management capabilities. Another cluster can be referred to as the commercial and environmental excellence group, which has a high level of pollution. This group has strongly developed environmental management capabilities and it perceives a high business network impact. Top management has given itself environmental responsibilities and wants to commercially exploit attention for the environment through environmental reputation enhancement and marketing opportunities. Environmental issues are dealt with by marketing and sales to commercially exploit environmentally friendly product (re)design.

Two groups have a medium pollution level, while the availability of environmental management capabilities differs significantly. A compliance-plus strategy can be recognized in the profile of the firms with strongly developed environmental management capabilities, since top management exposes environmental commitment under the influence of government and the local community (see Box 8.2). At the same time they perceive a remarkably low impact from chain actors, such as buyers and suppliers. The fact that their attention for the environment seems to have little to do with handling customer expectations implies a lack of commercial exploitation of their environmental activities. This is in line with the empirical findings, since these show that these firms perceive only limited benefits from environmental reputation enhancement and even less from environmental marketing opportunities (see Table 7.22). The other group has a medium environmental impact and weakly developed environmental management capabilities. They perceive, however, not only a strong impact from government, but also from suppliers, buyers, and consumers. Although top management shows environmental commitment, the perceived impact of the environmental manager is significantly lower compared to the compliance-plus and environmental excellence firms. These firms may be in the process of greening their business to the level of compliance-plus or commercial and environmental excellence based on satisfying environmental wishes of customers. Box 8.2 therefore refers to this group as 'environmental transition'.

In conclusion, the joint impact of the business network and firm characteristics appears to be different for the four cluster groups. The firms in the compliance and the commercial and environmental excellence clusters have found an adequate balance between stakeholder influences and firm strategy, on the one hand, and firm strategy and the development of environmental management capabilities, on the other. By contrast, the firms in the environmental transition cluster may have to change their business to commercial and environmental excellence to safeguard *license-to-produce*, while the environmental-plus firms seem to lack fit between their attention for the environment and a low perceived level of interests from buyers downwards in the chain.

Box 8.2. Environmental taxonomies.

Compliance-oriented (22% of the firms)
These firms have a limited environmental impact and focus on acting in compliance with regulative requirements. Their environmental management capabilities are poorly developed. Examples of these firms are medium-sized vegetables and fruit processors, which cause minor environmental pressures.

Environmental transition (34% of the firms)
These firms have a medium to high level of environmental impact. Although they are focused on compliance with environmental regulations, there is growing pressure from chain actors. Typically they perceive strong pressure from large and powerful suppliers, which want the firm to pay attention to the environment to ensure *license-to-produce* at the chain level. However, the development of environmental management capabilities is (still) lagging behind in these firms. Examples are medium-sized bakery factories.

Compliance-plus (25% of the firms)
These firms have a medium to high level of environmental impact. They typically perceive pressure from government, local inhabitants, and branch-organizations to clean up. They have well-developed environmental management capabilities. They have appointed an environmental manager to support the different departments with environmental activities. A lack of perceived environmental interests among actors downstream in the chain, such as customers and consumers, is felt as a barrier to commercially exploiting their environmental attention. Examples of these firms are medium-sized and large meat processors.

Commercial and environmental excellence (19% of the firms)
These firms have a medium to high level of environmental impact. They typically perceive a strong pressure from government, consumers, and bank and insurance firms to take care of the environment. The importance of developing environmental management capabilities is understood from a dynamic rather than a static perspective: they want to satisfy the environmental wishes of their customers on a proactive basis. They gain commercial benefits by an environmentally friendly reputation and they expect to gain environmental marketing opportunities. It is paramount that they do more than legally required based on the integration of environmental issues in all their business activities. They try to influence future environmental regulation and, as a way of showing environmental excellence to government, frequently participate in environmental voluntary agreements. Examples of these firms are prospectors with a strong focus on launching new products and using new production technologies (e.g. dairy processors and breweries).

Adapted from Roome (1994) and modified based on the cluster analysis.

8.6 Theoretical and methodological contribution

The theoretical framework integrates the inside-out and the outside-in perspective on the adoption of environmental management capabilities. It appears to be a fruitful approach to increase insight into the important factors that explain attention for the environment using theories on stakeholder influences, organizational resources, capabilities, and firm strategy (see Chapter 4). In doing so, different theoretical contributions can be discerned for the present study:

- It underpins the chain and network view that firms are part of a business network comprising multiple actors that can exert different types of influences on the firm. It is built on previous studies which adopt an outside-in perspective by investigating the business network impact (Madsen *et al.*, 1997; Henriques and Sadorsky, 1999; Braglia and Petroni, 2000; Madsen and Ulhøi, 2001). In addition, it evaluates the joint impact of the different stakeholders. The merit of this approach is shown in the fact that not only the relationship with government appears to significantly influence the availability of environmental management capabilities, but also other business network relationships, such as cooperation with buyers and/or suppliers as well as exerted pressure from suppliers, banks and insurance firms, and local inhabitants.

- It contributes to the limited quantitative empirical insight into the implications of voluntary public-private efforts on the attention for the environment (Sharma, 2001; Buysse and Verbeke, 2003). The results suggest that their implementation has a positive effect on the development of environmental management capabilities, although it should be noted that several interviewees indicated perceiving covenants as semi-enforced regulation.

- The empirical findings stress the importance of developing dynamic environmental capabilities to meet increased business network impacts encountered in the longitudinal analysis. As such, the present study contributes to the contingent-RBV, which emphasizes that competitiveness of organizational resources depends on market components, such as customers' interests (Aragón-Correa and Sharma, 2003). It is also noted that the empirical findings underpin the discussion by Hart (1995) on the natural-RBV, stating that competitiveness of environmental capabilities can be enhanced by consideration of environmental interests among stakeholders (Litz, 1996; Lober, 1998).

- The cluster analysis examines the joint effect of business network impacts, firm strategy, and organizational capabilities. It identifies profiles which show that firms are confronted with mismatches between stakeholder expectations and organizational capabilities. This insight contributes to the view of contingency scholars that business performance depends on the level of fit, congruence, or alignment between the internal and external business environment (Duncan, 1972; Hofer, 1975; Venkatraman, 1984; 1989).

A mixed research methodology is used to ensure validity and reliability of the empirical results. It consists of sequential triangulation of qualitative (i.e. semi-structured interviews) and quantitative (i.e. large-scale surveys) research strategies. The semi-structured interviews with environmental managers were essential to focus on the relevant issues to be included in the research framework. The large-scale surveys are used to investigate their importance in a larger sample of firms in the Dutch food and beverage industry. The interviews carried out after the surveys improved the interpretation of the empirical findings. In line with suggestions by other authors, the present study shows that qualitative and quantitative research strategies can be used in complementary ways (Foss and Ellefsen, 2002; Onwuegbuzie, 2002; Rallis and Rossman, 2003). Furthermore, insight into the empirical results benefited from the cluster analysis. It resulted in different organizational configurations, which can be defined as groups of firms that have the same profile of organizational characteristics (Meyer *et al.*, 1993; Ketchen *et al.*, 1997). The pooling of regression and cluster analysis as proposed by Ketchen and Shook (1996), using the significant predictors of the regression analysis as clustering variables, proved to be an effective strategy.

8.7 Suggestions for further research

Some limitations of the present study need to be discussed in the light of generalization of the results as well as suggestions for further research. The longitudinal analysis demonstrates significant increases in stakeholder influences between 2002 and 2005. Since the measurements are based on two points in time only, it might prevent the generalization of the longitudinal effects. It is therefore suggested that repeated measurements of business network impacts are carried out. Taking into account the results of this book, it is expected that the attention for the environment among chain actors has increased in particular in recent years. A longitudinal investigation is also relevant to get more insight into how firms align their business to changing stakeholder expectations. Strategic alignment is the dynamic process for finding and maintaining a profitable balance between firm strategy and stakeholder expectations, on the one hand, and firm strategy and the organizational capabilities on the other (Fortuin, 2006). It can also be remarked that the business network impacts are not measured in a longitudinal setting for micro and small firms. Parallel to the developments in medium-sized and large firms, the micro and small firms are probably also being confronted with increased stakeholder pressures. Finally, it would be interesting to further investigate the relationship between a prospector strategy and the development of cradle-to-cradle capabilities to reduce environmental impacts from a product life-cycle perspective.

8.8 Management and policy implications

Based on the results of the present study, management and policy implications will be drawn from the interdependency of firm strategy, business network influences, company size and sector. The management and policy implications are structured according to Figure 8.1.

Business strategy and company size (Arrow A, Figure 8.1)

For micro and small firms, strong ties with government through frequent contacts with local governmental agencies appear to be essential for staying informed about environmental

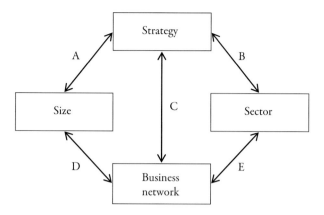

Figure 8.1. Interdependency between firm strategy, company size, sector and the business network (the arrows are discussed in the text).

regulations. The side effect of such regular contacts is that the understanding and consequently the appreciation of public environmental policy improves. Governmental agencies are therefore recommended to provide micro and small firms environmental information (e.g. through e-mails or news letters) and, more importantly, maintain frequency of visits even if no direct monitoring or control reasons exist. The micro and small firms typically lack the expertise to fundamentally change their environmental behavior. Environmental training and education programs could therefore be organized to increase the expertise to deal with environmental measures. Compliance-plus and environmental excellence firms adopt environmental measures beyond governmental requirements. They might best be included in environmental pilot projects to implement clean technologies to improve energy efficiency (e.g. using renewable energy and new cooling techniques) or to reduce waste production (e.g. using recycled and recyclable materials). Such cooperative projects can serve as platforms for knowledge exchange. It is important that intermediaries, such as branch-organizations, are involved in these initiatives, since they can take care of the administrative consequences. For micro and small firms environmental reporting on a company scale is too costly. To reduce the administrative burden, collective reporting, such as by the branch-organization, could significantly lower this burden and, in doing so, lower the barrier for transparent reporting on environmental issues. It could also be investigated as to when it is possible to base the environmental permits on the whole supply chain, rather than the individual firm. This might reduce administrative burden, while it focuses on the interconnections of different actors in the supply chain, stimulating attention for chain-oriented environmental management (Haverkamp *et al.*, 2005).

In comparison with the micro and small firms, the medium-sized and large firms have, in general, a larger budget available for environmental management activities. In the compliance-oriented firms the adoption of environmental management capabilities is most likely provoked by command-and-control regulation. Examples of these measures are compulsory emission reduction targets included in environmental permits, accompanied by strict monitoring to avoid non-compliance. The present study shows that firms with an innovative strategy (prospector strategy) are typically developing cradle-to-cradle capabilities (see Figure 7.3). To support this, government is advised to stimulate environmentally friendly product (re)design and production technologies, such as dissembling and recycling techniques. Furthermore, environmental covenants can put a strong focus on innovation addressing environmental targets at the product level (e.g. percentage of recycled components or re-usable parts) or at the chain level (e.g. chain level energy reduction targets). Large firms in particular are able to stimulate their chain partners to be included in environmental initiatives, which is essential to achieve chain-oriented environmental targets.

Business strategy and sector (Arrow B, Figure 8.1)

The interviews indicate that environmental managers perceive barriers to developing environmental management capabilities due to a lack of horizontal and vertical information exchange through intermediaries, such as branch-organizations (see Section 6.2.2). Intermediaries are therefore advised to put effort into the establishment of environmental communication channels by organizing environmental meetings, workshops and/or electronic discussion platforms. Firms and intermediaries should, however, be aware that overly tight relationships within the same sector (or branch) could also lead to 'lock-in' effects. Information exchange, including benchmarking with other sectors, may help to get valuable new insights into how to deal with environmental issues.

Business strategy and business network (Arrow C, Figure 8.1)

In this book, business network influences on environmental issues are evaluated beyond the traditionally important actors (i.e. government and local inhabitants), to include commercial stakeholders, such as buyers and suppliers, consumers, and banks and insurance firms. The growing significance of environmental criteria in financial services (e.g. loans and insurances) increases the importance for firms to show environmental excellence. As a consequence, environmental certification, such as ISO14001, might become increasingly important for guaranteeing environmental credibility towards banks and insurance firms. The attention for environmental issues at the chain level might lead to growing interest in environmentally friendly product (re)design as well. Seen in the light of the lagging development of cradle-to-cradle capabilities in the Dutch food and beverage industry, this suggests an important focus of attention. Firms should be aware that environmentally friendly product (re)design has numerous implications for business processes (Rocha and Brezet, 1999; De Bakker and Nijhof, 2002; Handfield et al., 2005). It requires close environmental cooperation of the different departments (e.g. R&D, production, but also purchasing, marketing and sales) for which open environmental communication channels are conditional. It is important that key persons in the organization, such as a board member and different department heads, get formal environmental responsibilities. The environmental manager should report directly to the firm's board and can provide support in terms of environmental training and education, developing environmental handbooks and carrying out internal audits to check conformity of business operations with the environmental goals.

Business network and company size and sector (Arrow D and E, Figure 8.1)

It should be emphasized that business network impacts differ between firms depending on company size and sector. The present study indicates that it is crucial for firms to constantly ask themselves whether their environmental activities are in line with the environmental wishes of government and local inhabitants. Anticipating governmental requirements can not only be done by gathering information via websites, professional journals, and/or branch-organizations, but also by consulting lower governmental civil servants to evaluate whether (planned) business operations are covered by the environmental permits. Local inhabitants can be involved in environmental issues by using their feedback in environmental complaints and information evenings to discuss the environmental situation of the firm. Additionally, medium-sized and large firms need to evaluate the environmental expectations of bank and insurance firms, as well as buyers, suppliers, and consumers to guarantee business continuation. These stakeholders will typically exert stronger environmental impacts on large compared to small firms as well as firms operating in sectors which cause substantive versus minor environmental pressures.

To ensure that enough attention is paid to the development of environmental management capabilities, managers are recommended to ask themselves the following questions (Clarke and Roome, 1999; Verheul, 1999; Prakash, 2001; Holt, 2004; Brammer and Pavelin, 2006; Marcus and Anderson, 2006):
- Is our firm putting more, equal, or less emphasis on environmental issues than other firms in the sector? Do we implement (enough) clean technologies? Do we put enough emphasis on environmentally friendly product (re)design? What does the implication of new environmental governmental and other requirements mean for our business operations? Do we need to implement new working routines to comply with the imposed environmental criteria? What are the advantages/disadvantages of entering environmental covenants, cooperative environmental projects and/or formal certification schemes, such as ISO14001? What can we learn from other firms in our sector that have implemented clean technologies?
- Are our chain partners active enough in controlling environmental emissions? And if not, how can we improve attention for the environment at the chain level?
- Is our branch-organization paying enough attention to the environment? Does it stimulate environmental initiatives through environmental brochures, information and workshops? Is it involved in the settlement of environmental agreements with government related to cradle-to-cradle aspects? If not, how can we stimulate our branch-organization to pay more attention to the environment?

- How does our sector perform in taking care of the environment in comparison with other sectors? Are examples known of sectors that perform better and can be used as benchmarks, such as with respect to participation in environmental covenants, achieving energy reduction targets, and developing cradle-to-cradle capabilities?

References

Adner, R. and C.E. Helfat, 2003. Corporate effects and dynamic managerial capabilities. Strategic Management Journal 24: 1011-1025.

Albrecht, J. and B. Arts, 2005. Climate policy convergence in Europe: an assessment based on National Communications to the UNFCCC. Journal of European Public Policy 12(5): 885-902.

Amit, R. and P.J.H. Schoemaker, 1993. Strategic assets and organizational rent. Strategic Management Journal 14: 33-46.

Aragón-Correa, J.A., 1998. Strategic proactivity and firm approach to the natural environment. Academy of Management Journal 41(5): 556-567.

Aragón-Correa, J.A. and S. Sharma, 2003. A contingent resource-based view of proactive corporate environmental strategy. Academy of Management Review 28(1): 71-88.

Armstrong, J.S. and T.S. Overton, 1977. Estimating nonresponse bias in mail surveys. Journal of Marketing Research 14(3): 396-402.

Backes, C.W. and N.J.M. Nelissen, 2000. Juridische sturing van milieuvraagstukken. Milieu, samenleving en beleid P.P.J. Driessen and P. Glasbergen. Den Haag, Elsevier: 281-303.

Banerjee, S.B., 2002. Corporate environmentalism: the construct and its measurement. Journal of Business Research 55: 177-191.

Bansal, P. and W.C. Bogner, 2002. Deciding on ISO 14001: economics, institutions, and context. Long Range Planning 35: 269-290.

Barney, J.B., 1986a. Organizational culture: can it be a source of sustained competitive advantage? Academy of Management Review 11(3): 656-665.

Barney, J.B., 1986b. Strategic factor markets: expectations, luck, and business strategy. Management Science 32(10): 1231-1241.

Barney, J.B., 2001. Is the resource based 'view' a useful perspective for strategic management research? Yes. Academy of Management Review 26(1): 41-56.

Bloemhof-Ruwaard, J.M., P. Van Beek, L. Hordijk and L.N. Van Wassenhove, 1995. Interactions between organizational research and environmental management. European Journal of Operational Research 85: 229-243.

Blomquist, T. and J. Sandström, 2004. From issue to checkpoints and back: managing green issues in R&D. Business Strategy and the Environment 13: 363-373.

Boiral, O., 2002. Tacit Knowledge and Environmental Management. Long Range Planning 35(3): 291-317.

Braakhuis, F.L.M., M. Gijtenbeek and W.A. Hafkamp (Eds.), 1995. Milieumanagement: van kosten naar baten. Alphen aan den Rijn, Samson H.D. Tjeenk Willink.

Braglia, M. and A. Petroni, 2000. Stakeholders' influence and internal championing of product stewardship in the Italian food packaging industry. Journal of Industrial Ecology 4(1): 75-92.

Brammer, S.J. and S. Pavelin, 2006. Corporate reputation and social performance: the importance of fit. Journal of Management Studies 43(3): 435-455.

References

Branzei, O., T.J. Ursacki-Bryant, I. Vertinsky and W. Zhang, 2004. The formation of green strategies in Chinese firms: matching corporate environmental response and individual principles. Strategic Management Journal 25: 1075-1095.

Brass, D.J. and M.E. Burkhardt, 1993. Potential power and power use: an investigation of structure and behavior. Academy of Management Journal 36(3): 441-470.

Bremmers, H.J., 2000. Milieuverslaggeving. The Hague, Elsevier business information.

Bremmers, H.J., S.W.F. Omta and M. Smit, 2003. Managing environmental information flows in food and agribusiness chains: a study on the relationship between ICT development and environmental performance. Wageningen, Wageningen University.

Brophy, M., 2004. Environmental guidelines and charters. Corporate environmental management: systems and strategies. R. Welford. London, EarthScan: 102-115.

Burt, R.S., 1992. Structural holes: the social structure of competition. Cambrdige, Massachusetts, London, Harvard University Press.

Buysse, K. and A. Verbeke, 2003. Proactive environmental strategies: a stakeholder management perspective. Strategic Management Journal 24: 453-470.

Carmines, E.G. and R.A. Zeller, 1979. Reliability and validity assessment. Bevery Hills/London, SAGE Publications.

Carroll, A.B., 1979. A three-dimensional conceptual model of corporate performance. Academy of Management Review 4(4): 497-505.

Christmann, P., 2000. Effects of 'best-practice' of environmental management on cost advantage: the role of complementary assets. Academy of Management Journal 43(4): 663-680.

Christopher, M.G., 1998. Logistics and supply chain management; strategies for reducing costs and improving services. London, Pitman Publishing.

Churchill, G.A., 1979. A paradigm for developing better measures of marketing constructs. Journal of Marketing Research 16: 64-73.

Churchill, G.A., 1999. Marketing research: methodological foundations. Orlando, The Dryden Press.

Clarke, S. and N. Roome, 1999. Sustainable business: learning - action networks as organizational assets. Business Strategy and the Environment 8: 296-310.

Clarke, S.F., 1998. Sustainable technology management: the role of networks of learning. Sustainable strategies for industry. N.J. Roome. Washington DC, Island Press.

Clarkson, M.B.E., 1991. Defining, evaluating and managing corporate social performance: The stakeholder management model. Research in corporate social performance and policy. L.E. Preston. Greenwich, CT, JAI Press. 12: 331-358.

Clarkson, M.B.E., 1995. A stakeholder framework for analyzing and evaluating corporate social performance. The Academy of Management Review 20(1): 92-117.

Claro, D.P., 2003. Managing business networks and buyer-supplier relationships: how information obtained from the business network affects trust, transaction specific investments, collaboration and performance in the Dutch potted plant and flower industry (Dissertation). Business Administration. Wageningen, Wageningen University.

Clemens, B.W. and T.J. Douglas, 2006. Does coercion drive firms to adopt 'voluntary' green initiatives? Relationships among coercion, superior firm resources, and voluntary green initiatives. Journal of Business Research 59(4): 483-491.

Coase, R.H., 1937. The nature of the firm. Economica 4: 386-405.

Coleman, J.S., 1988. Social capital in the creation of human capital. American Journal of Sociology 94(Supplement: Organizations and Institutions: Sociological and Economic Approaches to the Analysis of Social Structure): S95-S120.

Coleman, J.S., 1990. Foundations of social theories. Cambridge, MA, Harvard University Press.

Cramer, J., 1998. Environmental management: from 'fit' to 'stretch'. Business Strategy and the Environment 7: 162-172.

Cramer, J., 2005. Company learning about corporate social responsibility. Business Strategy and the Environment 14: 255-266.

Cramer, J., R. Kim and E. van Dam, 2004. Corporate social responsibility in Dutch industry. Corporate Social Responsibility and Environmental Management 11: 188-195.

Dankaart, D. 2007. Tripod Beta - De vergeten stap... Veiligheid, waar ligt de grens? 1947-2007: een wereld van verschil, Arnhem, Nederlandse Vereniging voor Veiligheidskunde.

Day, G.S., 1994. The capabilities of market-driven organizations. Journal of Marketing 58(4): 37-52.

De Bakker, F. and A. Nijhof, 2002. Responsible chain management: a capability assessment framework. Business Strategy and the Environment 11: 63-75.

De Leeuw, A.C.J., 2000. Bedrijfskundig management: primair proces, strategie, en organisatie. Assen, Van Gorcum.

Del Brío, J.A. and B. Junquera, 2003. A Review of the literature on environmental innovation management in SMEs: implications for public policies. Technovation 23: 939-948.

Diamantopoulos, A., 1999. Export performance measurements: reflective versus formative indicators. International Marketing Review 16(6): 444-457.

Diamantopoulos, A. and H.M. Winklhofer, 2001. Index construction with formative indicators: an alternative to scale development. Journal of Marketing Research XXXVIII(May): 269-277.

Dierickx, I. and K. Cool, 1989. Asset stock accumulation and sustainability of competitive advantage. Management Science 35(12): 1504-1511.

Dillman, D.A., E. Carpenter, J. Christensen and R. Brooks, 1974. Increasing mail questionnaire response: a four state comparison. American Sociological Review 39(5): 744-756.

Donaldson, T. and L.E. Preston, 1995. The stakeholder theory of the corporation: concepts, evidence, and implications. Academy of Management Review 20(1): 65-91.

Driscoll, C. and M. Starik, 2004. The primordial stakeholder: advancing the conceptual consideration of stakeholder status for the natural environment. Journal of Business Ethics 49: 55-73.

Duncan, R.B., 1972. Characteristics of organizational environments and perceived environmental uncertainty. Administrative Science Quarterly 17(3): 313-327.

Dutilh, C.E. and C. Blijswijk, 2004. Milieu krijgt steeds meer aandacht. VMT(18-19): 12-14.

Eesly, C. and M.J. Lenox, 2006. Firm responses to secondary stakeholder action. Strategic Management Journal 27: 765-781.

Eisenhardt, K.M., 1989. Building theories from case study research. Academy of Management Review 14(4): 532-550.

Eisenhardt, K.M. and J.A. Martin, 2000. Dynamic capabilities: what are they? Strategic Management Journal 21: 1105-1121.

References

Elkington, J., 1998. Cannibals with forks: the triple bottom line of 21st century business. Gabriola Island BC, New Society Publishers.

Epstein, M.J. and M.-J. Roy, 2001. Sustainability in action: identifying and measuring the key performance drivers. Long Range Planning 34: 585-604.

European-Commission, 2001. Promoting a European framework for corporate social responsibility (green paper), Directorate-general for employment and social affairs.

Field, A., 2003. Discovering Statistics using SPSS for Windows. London, SAGE Publications Ltd.

Fortuin, F., 2006. Aligning innovation to business strategy: combining cross-industry and longitudinal perspectives on strategic alignment in leading technology-based companies (Dissertation). Wageningen, Wageningen University.

Foss, C. and B. Ellefsen, 2002. The value of combining qualitative and quantitative approaches in nursing research by means of method triangulation. Journal of Advanced Nursing 40(2): 242-248.

Fowler, F.J., 2002. Survey research methods. Thousand Oaks/London, Sage Publications Ltd.

Fredericks, E., 2005. Infusing flexibility into business-to-business firms: a contingency theory and resource-based view perspective and practical implications. Industrial Marketing Management 34: 555-565.

Freeman, R.E., 1984. Strategic management: A stakeholder approach. Boston, Pitman/Ballinger.

Friedman, H.H. and T. Amoo, 1999. Rating the rating scales. Journal of Marketing Management 9(3): 114-123.

Frooman, J., 1999. Stakeholder influence strategies. The Academy of Management Journal 24(2): 191-205.

Fryxall, G.E. and M. Vryza, 1999. Managing environmental issues across multiple functions: an empirical study of corporate environmental departments and functional co-ordination. Journal of Environmental Management 55: 39-56.

Fryxell, G.E., S.S. Ching and C.W.H. Lo, 2004. Does the selection of ISO 14001 registrars matter? Registrar reputation and environmental policy statements in China. Journal of Environmental Management 71: 45-57.

Fryxell, G.E. and A. Szetto, 2002. The influence of motivations for seeking ISO 14001 certification: an empirical study of ISO 14001 certified facilities in Hong Kong. Journal of Environmental Management 65: 223-238.

Garrod, B. and P. Chadwick, 1996. Environmental management and business strategy: towards a new strategic paradigm. Futures 28(1): 37-50.

Geelhoed, L.A., 1992. Wetgeving, de rol in het milieubeleid. Milieu(2): 57-62.

Gerrard, B., 1990. On matters methodological in economics. Journal of economic surveys 42(2): 197-219.

Ghobadian, A., H. Viney, J. Liu and P. James, 1998. Extending linear approaches to mapping corporate environmental behaviour. Business Strategy and the Environment 7: 13-23.

Gladwin, T.N., J.J. Kennelly and T.S. Krause, 1995. Shifting paradigms for sustainable development: implications for management theory and research. The Academy of Management Review 20(4): 874-907.

Glasbergen, P., 1999. Tailor-made environmental governance: on the relevance of the covenanting process. European Environment 9: 49-58.

Glasbergen, P. and P.P.J. Driessen, 2002. The paradigm shift in environmental politics: towards a new image of the manageable society. Green society: the paradigm shift in Dutch environmental politics. P.P.J. Driessen and P. Glasbergen. Dordrecht, Boston, London, Kluwer Academic Publishers: 3-25.

González-Benito, J. and O. González-Benito, 2005. Environmental proactivity and business performance: an empirical analysis. Omega 33: 1-15.

Govindarajulu, N. and B.F. Daily, 2004. Motivating employees for environmental improvement. Industrial Management & Data Systems 104(4): 364-372.

Granovetter, M., 1973. The strength of weak ties. The American Journal of Sociology 78(6): 1360-1380.

Granovetter, M., 1985. Economic action and social structure: the problem of embeddedness. The American Journal of Sociology 91(3): 481-510.

Granovetter, M., 1992. Problems of explanation in economic sociology. Handbook of economic sociology. N. Nohria and R.G. Eccles. Boston, Harvard Business School Press: 25-56.

Gupta, M.C., 1994. Environmental management and its impact on the operations function. International Journal of Operations & Production Management 15(8): 34-51.

Hahn, T. and M. Scheermesser, 2006. Approaches to corporate sustainability among German companies. Corporate Social Responsibility and Environmental Management 13: 150-165.

Hair, J.F., R.E. Anderson, R.L. Tatham and W.C. Black, 1998. Multivariate data analyses. Upper Saddle River, New Jersey, Prentice-Hall, Inc.

Hajer, M.A., 1995. The politics of environmental discourse: ecological modernization and policy process. Oxford, Clarendon Press.

Håkansson, 1982. International marketing and purchasing of industrial goods. Chichester, Jonhn Wiley & Sons.

Håkansson, H. and I. Snehota, 1989. No business is an island: the network concept of business strategy. Scandinavian Journal of Management 5(3): 187-200.

Halkos, G.E. and K.I. Evangelinos, 2002. Determinants of environmental management systems standards implementation: evidence from Greek industry. Business Strategy and the Environment 11: 360-375.

Handfield, R., R. Sroufe and S. Walton, 2005. Integrating environmental management and supply chain strategies. Business Strategy and the Environment 14: 1-19.

Hart, S.L., 1995. A natural-Resource-Based View of the firm. The Academy of Management Review 20(4): 986-1014.

Hart, S.L., 1997. Beyond greening: strategies for a sustainable world. Harvard Business Review Jan-Feb: 67-76.

Haverkamp, D.J., H.J. Bremmers and M. De Rooij, 2005. Milieuzorg geketend of milieuzorg in ketens? Bestuurskunde 14(5): 29-38.

Hedstrom, G., J. Keeble, D. Lyon, A. Pardo and D. Vassallo, 2005. The innovation high ground: how leading companies are using sustainability-driven innovation. PRISM(1): 9-27.

Henriques, I. and P. Sadorsky, 1996. The determinants of an environmentally responsive firm: an empirical approach. Journal of Environmental Economics and Management 30(3): 381-395.

References

Henriques, I. and P. Sadorsky, 1999. The relationship between environmental commitment and managerial perceptions of stakeholder importance. Academy of Management Journal 42(1): 87-99.

Hillary, R., 1999. Evaluation of study reports on the barriers, opportunities and drivers for small and medium sized enterprises in the adoption of environmental management systems. London, Network for environmental management and accounting.

Hillary, R., 2004. Environmental management systems and the smaller enterprise. Journal of Cleaner Production 12: 561-569.

Hobbs, J.E., 1996. A transaction cost approach to supply chain management. Supply Chain Management 1(2): 15-27.

Hofer, C.W., 1975. Toward a contingency theory of business strategy. The Academy of Management Journal 18(4): 784-810.

Holmberg, J. and K.H. Robèrt, 2000. Backcasting from non-overlapping sustainability principles - a framework for strategic planning. International Journal of Development and World Ecology 7(291-308).

Holt, D., 2004. Managing the interface between suppliers and organizations for environmental responsibility - an exploration of current practices in the UK. Business Strategy and the Environment 11: 71-84.

Huizing, A., 1993. Alternatieve vormen van milieujaarverslagleving. Pacioli journaal(Maart): 10-13.

James, P., 1994. Business environmental performance measurement. Business Strategy and the Environment Summer: 59-67.

Jarvis, C.B., S.B. MacKenzie and P.M. Podsakoff, 2003. A critical review of construct indicators and measurement model misspecification in marketing and consumer research. Journal of Consumer Research 30: 199-218.

Jordan, A. and D. Liefferink (Eds.), 2004. Environmental policy in Europe: the Europeanization of national environmental policy. London, New York, Routledge.

Judge, W.Q. and T.J. Douglas, 1998. Performance implications of incorporating natural environmental issues into the strategic planning process: an empirical assessment. Journal of Management Studies 35(2): 241-262.

Jurgens, G.T.J.M. and M.A. Wiering, 2000. Handhaving van milieuregels. Milieu, samenleving en beleid. P.P.J. Driessen and P. Glasbergen. Den Haag, Elsevier: 305-326.

Kagan, R.A., N. Gunningham and D. Thornton, 2003. Explaining corporate environmental performance: how does regulation matter? Law & Society Review 37(1): 51-89.

Kaplan, B. and D. Duchon, 1988. Combining qualitative and quantitative methods in information systems research: a case study. MIS Quarterly 12(4): 571-586.

Karapetrovic, S. and W. Willborn, 1998. Integration of quality and environmental management systems. The TQM Magazine 10(3): 204-213.

Ketchen Jr., D.J., J.G. Combs, C.J. Russell, C. Shook, M.A. Dean, J. Runge, F.T. Lohrke, S.E. Naumann, D.E. Haptonstahl, R. Baker, B.A. Beckstein, C. Handler, H. Honig and S. Lamoureux, 1997. Organizational configurations and performance: a meta-analysis. Academy of Management Journal 40(1): 223-240.

Ketchen Jr., D.J. and C.L. Shook, 1996. The application of cluster analysis in strategic management research: an analysis and critique. Strategic Management Journal 17(6): 441-458.

King, A., 2000. Organizational response to environmental regulation: punctuated change or autogenesis. Business Strategy and the Environment 9: 224-238.

Klassen, R.D. and C.P. McLaughlin, 1996. The impact of environmental management on firm performance. Management Science 42(8): 1199-1214.

Klassen, R.D. and D.C. Whybark, 1999. The impact of environmental technologies on manufacturing performance. Academy of Management Journal 42(6): 599-615.

Kolk, A., 2000. Economics of environmental management. Essex, Pearson Education Limited.

Kolk, A. and A. Mauser, 2002. The evolution of environmental management: from stage models to performance evaluation. Business Strategy and the Environment 11: 14-31.

Lambert, D.M. and M.C. Cooper, 2000. Issues in supply chain management. Industrial marketing management 29: 65-83.

Lazzarini, S.G., F.R. Chaddad and M.L. Cook, 2001. Integrating supply chain and network analysis: the study of netchains. Chain and network science 1(1): 7-22.

Lee, K.H. and R. Ball, 2003. Achieving sustainable corporate competitiveness: strategic link between top management 's (green) commitment and corporate environmental strategy. Greener Management International 44(Winter): 89-104.

LEI, 2003. Landbouweconomisch bericht 2003. P. Berkhout and C. Van Bruchem. Den Haag, Landbouweconomisch instituut.

LEI, 2004. Landbouweconomisch bericht 2004. P. Berkhout and C. Van Bruchem. Den Haag, Landbouweconomisch instituut.

LEI, 2005. Landbouw economisch bericht 2005. P. Berkhout and C. Van Bruchem. Den Haag, Landbouw Economisch Instituut.

LEI, 2006. Landbouw-economisch bericht 2006. P. Berkhout and C. Van Bruchem. Den Haag, Landbouweconomische instituut: 228.

Leonard-Barton, D., 1992. Core capabilities and core rigidities: a paradox in managing new product development. Strategic Management Journal 13: 111-125.

Lippmann, S., 1999. Supply chain environmental management: elements for success. Environmental Management 6(2): 175-182.

Litz, R.A., 1996. A resource-based-view of the socially responsible firm: stakeholder interdependence, ethical awareness, and issue responsiveness as strategic assets. Journal of Business Ethics 15: 1355-1363.

LNV, 2005. Feiten en cijfers van de Nederlandse agrosector 2005/2006. Den Haag, Ministerie van Lanbouw, Natuur en Voedselkwaliteit.

Lober, D.J., 1998. Pollution prevention as corporate entrepreneurship. Journal of Organizational Change 11(1): 26-37.

Madsen, H., K. Sinding and J.P. Ulhøi, 1997. Sustainability and corporate environmental focus: an analysis of Danish small and medium sized companies. Managerial and Decision Economics 18(6): 443-453.

Madsen, H. and J.P. Ulhøi, 2001. Integrating environmental and stakeholder management. Business Strategy and the Environment 10: 77-88.

References

Marcus, A.A. and M.H. Anderson, 2006. A general dynamic capability: does it propagate business and social competencies in the retail food industry. Journal of Management Studies 43(1): 19-46.

Markides, C.C. and P.J. Williamson, 1994. Related diversification, core competences and corporate performance. Strategic Management Journal 15: 149-165.

Markides, C.C. and P.J. Williamson, 1996. Corporate diversification and organizational structure: a resource-based view. The Academy of Management Journal 39(2): 340-367.

Martin, R., 1998. ISO 14001 guidance manual. Tennessee, National Center for Environmental Decision-Making Research.

McDonough, W. and M. Braungart, 2002. Cradle to cradle: remaking the way me make things. New York, North Point.

McKinsey, 1991. The corporate response to the envrionmental challenge. Amsterdam, McKinsey.

McVea, J.F. and R.E. Freeman, 2005. A names-and-faces approach to stakeholder management: how focussing on stakeholders as individuals can bring ethic and entrepreneurial strategy together. Journal of management Inquiry 14(1): 57-69.

Melnyk, S.A., R.P. Sroufe and R. Calantone, 2003. Assessing the impact of environmental management systems on corporate and environmental performance. Journal of Operations Management 21(3): 329-351.

Menguc, B. and L.K. Ozanne, 2005. Challenges of the 'green imperative': a natural resource-based approach to the environment orientation-business performance relationship. Journal of Business Research 58(4): 430-438.

Meyer, A.D., A.S. Tsui and C.R. Hinings, 1993. Configurational approaches to organizational analysis. The Academy of Management Journal 36(66): 1175-1195.

Miles, R.E., C.C. Snow, A.D. Meyer and H.J. Coleman Jr., 1978. Organizational Strategy, Structure, and Process. Academy of Management Review 3(3): 546-562.

Miles, R.E. and E.E. Snow, 1978. Organizational strategy, structure, and process. New York, McGraw Hill.

Milieudefensie, 2007. Investing in climate change Dutch banks compared. Amsterdam, Friends of the Earth Netherlands.

Milligan, G.W., 1980. An examination of the effect of six types of error perturbation on fifteen clustering algorithms. Psychometrika 45(3): 325-342.

Mitchell, R.K., B.R. Agle and D.J. Wood, 1997. Toward a theory of stakeholder identification and salience: defining the principle of who and what really counts. Academy of Management Review 22(4): 853-886.

Mol, A.P.J., V. Lauber and D. Liefferink (Eds.), 2000. The voluntary approach to environmental policy: joint environmental approach to environmental policy-making in Europe. Oxford, Oxford University Press.

Mol, G., S.P. Vriend and P.F.M. Van Gaans, 2001. Environmental monitoring in the Netherlands: past developments and future challenges. Environmental Monitoring and Assessment(68): 313-335.

Morse, J., 1991. Approaches to qualitative-quantitative methodological triangulation. Nursing Research 40: 120-123.

Nelson, R.R. and S.G. Winter, 1982. An evolutionairy theory of economic change. Cambridge, Mass, Belknap Press.

Netherwood, A., 2004. Environmental management systems. Corporate environmental management: systems and strategies. R. Welford. London, EarthScan: 37-60.

Newton, T. and G. Harte, 1997. Green business: technicist kitsch? Journal of Management Studies 34(1): 75-98.

NVC, 2003. De Nederlandse verpakkingsstatistiek 2003. Gouda, Nederlands Verpakkingscentrum.

NZO/NEDSMELT, 2005. Sectormeerjarenplan-milieu zuivelindustrie 2005-2008. Zoetermeer, Nederlandse Zuivel Organisatie (NZO) en NEDSMELT: 9.

Oliver, C., 1991. Strategic responses to institutional processes. Academy of Management Review 16(1): 145-179.

Omta, S.W.F. and P. Folstar, 2005. Integration of innovation in the corporate strategy of agri-food companies. Innovation in agri-food systems. W.M.F. Jongen and M.T.G. Meulenberg. Wageningen, Wageningen Academic Publishers.

Omta, S.W.F., J.H. Trienekens and G. Beers, 2001. Chain and network science: a research framework. Journal on Chain and Network Science 1(1): 1-6.

Onwuegbuzie, A.J., 2002. Positivists, post-positivists, post-structuralists, and post-modernists: why can't we all get along? Towards a framework for unifying research paradigms. Education 122: 518-530.

Onwuegbuzie, A.J. and L.G. Daniel, 2003. Typology of analytical and interpretational errors in quantitative and qualitative educational research. Current Issues in Education 6(2).

Pavitt, K., 1984. Sectoral patterns of technical change: towards a taxonomy and a theory. Research Policy 13: 343-373.

PCCC, 2006. De staat van het klimaat 2006: actueel onderzoek en beleid nader verklaard, Platform Communication on Climate Change: 23.

Peattie, K. and A. Crane, 2005. Green marketing: legend, myth, farce or prophesy? Qualitative market research: an international journal 8(4): 357-370.

Penrose, E., 1959. The theory of the growth of the firm. New York, Wiley.

Pfeffer, J. and G.R. Salancik, 1978. The external control of organizations. New York, Harper & Row.

Polonsky, M.J., 1995. A stakeholder theory approach to designing environmental marketing strategy. Journal of business & industrial marketing 10(3): 29-46.

Porter, M.E., 1980. Competitive strategy: techniques for analyzing industries and competitors. New York, Free Press.

Porter, M.E., 1981. The contributions of industrial organization to strategic management. Academy of Management Review 6(4): 609-620.

Porter, M.E. and M.R. Kramer, 2006. Strategy & society: the link between competitive advantage and corporate social responsibility. Harvard Business Review(December): 78-92.

Porter, M.E. and C. Van der Linde, 1995. Green and competitive. Harvard Business Review 73: 120-134.

Powell, W.W., 1990. Neither market nor hierachy: network forms of organization. Research in Organizational Behavior 12: 295-336.

Prahalad, C.K. and G. Hamel, 1990. The core competence of the corporation. Harvard Business Review May-June: 79-91.

References

Prakash, A., 2000. Greening the firm. Cambridge, New York, Melbourne, Cambridge University Press.

Prakash, A., 2001. Why do firms adopt 'beyond-compliance' environmental policies? Business Strategy and the Environment 10: 286-299.

Prakash, A., 2002. Green marketing, public policy and managerial strategies. Business Strategy and the Environment 11: 285-297.

Prakash, A. and K. Kollman, 2004. Policy modes, firms and the natural environment. Business Strategy and the Environment 13: 107-128.

Priem, R.L. and J.E. Butler, 2001a. Is the resource-based 'view' a useful perspective for strategic management research? Academy of Management Review 26(1): 22-40.

Priem, R.L. and J.E. Butler, 2001b. Tautology in the resource-based view and the implications of externally determined resource value: further comments. Academy of Management Review 26(1): 57-66.

Quazi, H.A., Y.-K. Khoo, C.-M. Tan and P.-S. Wong, 2001. Motivation for ISO 14000 certification: development of a predictive model. Omega 29(6): 525.

Rallis, S.F. and G.B. Rossman, 2003. Mixed methods in evaluation contexts: a pragmatic framework. Handbook of mixed methods in social & behavorial research. A. Tashakkori and C. Teddlie. London, SAGE: 491-512.

Raskin, J.D., 2002. Constructivism in psychology: personal construct psychology, radical constructivism, and social constructionism. Studies in meaning: exploring constructivist psychology. J.D. Raskin and S.K. Bridges. New York, Pace University Press: 1-25.

Rhee, S.K. and S.Y. Lee, 2003. Dynamic change of corporate environmental strategy: rhetoric and reality. Business Strategy and the Environment 12: 175-190.

RMK, 1997. Evaluatie bedrijfsmilieuzorgsystemen 1996. The Hague, Raad voor Midden- en Kleinbedrijf.

Robèrt, K.H., B. Schmidt-Bleek, J. Aloisi de Larderel, G. Basile, J.L. Jansen, R. Kuehr, P. Price Thomas, M. Suzuki, P. Hawken and M. Wackernagel, 2002. Strategic sustainable development - selection, design and synergies of applied tools. Journal of Cleaner Production 10(3): 197-214.

Rocha, C. and H. Brezet, 1999. Product-oriented environmental management systems: a case study. The Journal of Sustainable Product Design (July): 30-42.

Roome, N., 1994. Business strategy, R&D management and environmental imperatives. R&D Management 24: 65-81.

Root, T.L., J.T. Price, R.H. Kimberly, S.H. Schneiders, C. Rosenzweig and A. Pounds, 2003. 'Fingerprints' of global warming on wilde animals and plants. Nature 421: 57-60.

Rowley, T.J., 1997. Moving beyond dyadic ties: a network theory of stakeholder influences. Academy of Management Review 22(4): 887-910.

Roy, M.-J., O. Boiral and D. Lagacé, 2001. Environmental commitment and manufacturing excellence: a comparative study within Canadian industry. Business Strategy and the Environment 10: 257-268.

Rumelt, R.P., 1991. How much does industry matter? Strategic Management Journal 12(3): 167-185.

Russo, M.V. and P.A. Fouts, 1997. A resource-based perspective on corporate environmental performance and profitability. Academy of Management Journal 40(3): 534-559.

Sanchez, R., 2001. Building blocks for strategy theory: resources, dynamic capabilities and competences. Rethinking strategy. H.W. Volberda and T. Elfring. London, Sage Publications: 143-158.

Sanchez, R., 2004. Understanding competence-based management: identifying and managing five models of competence. Journal of Business Research 57: 518-532.

Schaltegger, S., K. Müller and H. Hindrichsen, 1996. Corporate Environmental Accounting. Chichester, John Wiley & Sons.

Scholten, V.E., 2006. The early growth of academic spin-offs: factors influencing the early growth of Dutch spin-offs in the life sciences, ICT and consulting (dissertation). Business Administration. Wageningen, Wageningen University.

Sechrest, L. and S. Sidani, 1995. Quantitative and qualitative methods: is there an alternative. Evaluation and program planning 18(1): 77087.

SenterNovem, 2005. Energie uit uw reststromen, het kan! Praktijkvoorbeelden bio-energie voor de voedingsmiddelenindustrie, (Publicatienummer 2DEN-05.13).

Sharma, S., 2001. Different strokes: regulatory styles and environmental strategy in the North-American oil and gas industry. Business Strategy and the Environment 10: 344-364.

Sharma, S. and I. Henriques, 2005. Stakeholder influences on sustainability practices in the Canadian forst products industry. Strategic Management Journal 26: 159-180.

Sharma, S. and H. Vredenburg, 1998. Proactive corporate environmental strategy and the development of competitively valuable organizational capabilities. Strategic Management Journal 19(8): 729-753.

Shrivastava, P., 1995. Environmental technologies and competitive advantage. Strategic Management Journal 16: 183-200.

Sieber, S.D., 1973. The integration of fieldwork and survey methods. The American Journal of Sociology 78(6): 1335-1359.

Sinclair, D., 1997. Self-regulation versus command and control? Beyond false dichotomies. Law & Policy 19(4): 529-559.

Slack, N., S. Chambers and R. Johnston, 2001. Operations management. New York, Prentice Hall.

Sroufe, R., 2003. Effects of environmental management systems on environmental management practices and operations. Production and operations management 12(3): 416-431.

Starkey, R., 2004. The standardization of environmental systems: ISO 14001, ISO 14004 and EMAS. Corporate environmental management 1: systems and strategies. R. Welford. London, Earthscan: 61-101.

Stocké, V., 2006. Attitudes toward surveys, attitude accessibility and the effect on respodent's susceptibility to nonresponse. Quality & quantity 40: 259-288.

Subramaniam, M. and M.A. Youndt, 2005. The influence of intellectual capital on the types of innovative capabilities. Academy of Management Journal 48(3): 450-463.

Suchman, M.C., 1995. Managing legitimacy: strategic and institutional approaches. Academy of Management Review 20(3): 571-610.

Teece, D.J., 1986. Profiting from technological innovation: implications for integration, collaboration, licensing and public policy. Research policy 15: 285-305.

Teece, D.J., G. Pisano and A. Shuen, 1997. Dynamic capabilities and strategic management. Strategic Management Journal 18(7): 509-533.

UNCTAD, 1993. Environmental management in transnational corporations: report on the benchmark corporate environmental survey. New York, United Nations.

Uzzi, B., 1996. The sources and consequences of embeddedness for the economic performance of organizations: the network effect. American Sociological Review 61(4): 674-698.

Uzzi, B., 1997. Social structure and competition in interfirm networks: the paradox of embeddedness. Administrative Science Quarterly 42(1): 35-67.

Van der Kolk, J., 1987. Systematische milieuzorg in bedrijven. Milieu en recht(2): 42-48.

Van Snellenberg, T. and R. Van de Peppel, 2002. Perspectives on compliance: non-compliance with environmental licences in the Netherlands. European Environment 12: 131-148.

Van Tulder, R. and A. Van der Zwart, 2006. International business-society management: linking corporate responsibility and globalization. London and New York, Routledge.

Venkatraman, N., 1984. Exploring the concept of 'fit' in strategic management. Academy of Management Review 9(3): 513-525.

Venkatraman, N., 1989. The concept of fit in strategy research: toward verbal and statistical correspondence. Academy of Management Review 14(3): 423-444.

Verheul, H., 1999. How social networks influence the dissemination of cleaner technologies to SMEs. Journal of Cleaner Production 7(3): 213-219.

Vermeulen, W.J.V., 2002. Greening production as co-responsibility. Greening society: the paradigm shift in Dutch environmental politics. P.P.J. Driessen and P. Glasbergen. Dordrecht, Boston, London, Kluwer Academic Publishers: 67-90.

VNCI, 2003. Product stewardship for marketeers. Leidschendam, Association of Dutch Chemical Industry.

VROM, 2005. Hoofdlijnennotitie Toekomstagenda Milieu. The Hague, Ministry of VROM.

VROM, 2006a. Future Environmental Agenda: clean, clever, competetive. Ministry of Housing, Spatial Planning, and the Environment (VROM). The Hague.

VROM, 2006b. Informatieblad beheer verpakkingen en papier en karton. A. Directie Stoffen, Straling.

Walley, N. and B. Whitehead, 1994. It's not easy being green. Harvard Business Review May-June: 46-52.

Wätzold, F., A. Bültmann, M. Eames, K. Lulofs and S. Schicht, 2001. EMAS and regulatory relief in Europe: lessons from national experience. European Environment 11: 37-48.

WCED, 1987. Our common future. Oxford, Oxford university press.

Weber, M., 1947. The theory of social and economic organization. New York, Free Press.

Welford, R., 2004. Corporate environmental management 1: systems and strategies. London, EarthScan.

Welford, R.J., 1998. Corporate environmental management, technology and sustainable development: postmodern perspectives and the need for a critical research agenda. Business Strategy and the Environment 7: 1-12.

Welford, R.J., 2000. Corporate environmental management 3: towards sustainable development. London, Earthscan Publications Ltd.

Wernerfelt, B., 1984. A resource-based view of the firm. Strategic Management Journal 5(2): 171-180.

Wilkinson, G. and B.G. Dale, 1999. Integrated management systems: an examination of the concept and theory. The TQM Magazine 11(2): 95-104.

Williamson, O.E., 1985. The economic institutions of capitalism. New York, Free Press.

Yin, R.K., 2003. Case study research: design and methods. California, London, New Delhi, Sage Publications.

Zajac, E.J., M.S. Kraatz and K.F. Bresser, 2000. Modeling the dynamics of strategic fit: a normative approach to strategic change. Strategic Management Journal 21: 429-453.

Zobel, T. and J.O. Burman, 2004. Factors of importance in identification and assessment of environmental aspects in an EMS context: experiences in Swedish organizations. Journal of Cleaner Production 12: 13-27.

Zutshi, A. and A. Sohal, 2004a. Environmental management system adoption by Australasian organisations: part 1: reasons, benefits and impediments. Technovation 24(4): 335-357.

Zutshi, A. and A. Sohal, 2004b. A study of the environmental management system (EMS) adoption process within Australasian organisations - 2. Role of stakeholders. Technovation 24: 371-386.

Appendices

Appendix A. Interview protocol

A. Respondent's profile
1. Function of the respondent
2. Time spent in the firm and working in current function
3. Main activities of the respondent

B. The firm
1. Name and date of establishment
2. Core business and products
3. Turnover and number of employees (2001, 2004, and expected 2007)
4. Most important buyers and suppliers
5. Which business operations cause a (great) pressure on the environment?
6. Is your firm ISO14001-certified?

C. Organizational environmental characteristics
1. What are the main responsibilities of the environmentally responsible manager?
2. How big is his/her influence on various operational and strategic environmental activities? Can you give recent examples?
3. How are employees stimulated to share their environmental experience and ideas with each other and higher-level management?
4. What environmental issues are considered for environmental improvement of production processes and/or the development of new products?
5. To what extent are environmental issues integrated in food quality and safety systems or other management systems?
6. Does environmental management contribute to the corporate goals?

D. Business network impacts and relationships
1. What are the most important environmental issues that are dealt with in the relationship with (lower) governmental agencies?
2. How is your firm dealing with proposed and/or expected future environmental governmental regulations?
3. Does your firm receive environmental support from another chain actor and/or is your firm providing help to other chain actors?
4. Are environmental agreements made with buyers and/or suppliers? What environmental issues are dealt with in these agreements and how are they evaluated?
5. How is pressure from societal groups to reduce environmental impact dealt with?

E. Environmental strategy and performance

1. What were the most important reasons or causes of environmental problems that occurred over the past three years and how were they taken care of?
2. What environmental costs and benefits are perceived at the different management levels?
3. What are the most important motives for paying more attention to the environment other than those that are legally required?

Appendix B. Questionnaire 2002

1. What is your main position in the organization?

☐ Environmental Coordinator ☐ CEO
☐ Quality Manager ☐ Heath & safety manager
☐ Controller ☐ Product Manager
☐ Board Member ☐ Other: ....

2. Please indicate to what extent environmental pollution is a focus of attention in the organization

Not at all –   1   2   3   4   5   – Very much

1	Soil pollution	1	2	3	4	5
2	Noise	1	2	3	4	5
3	Water pollution	1	2	3	4	5
4	Smell	1	2	3	4	5
5	Air pollution	1	2	3	4	5
6	Hazardous substances	1	2	3	4	5
7	Waste production	1	2	3	4	5

3. Please indicate the reasons for collecting environmental information

1 ☐ Internal environmental management
2 ☐ Product (re)design
3 ☐ Chain-oriented environmental management
4 ☐ To check environmental emissions
5 ☐ To evaluate environmental measures
6 ☐ To exchange information with buyers/ suppliers

4. Please indicate the availability of the following elements of an environmental management system

1 ☐ Environmental strategy
2 ☐ Environmental audit
3 ☐ Environmental action program
4 ☐ Environmental training of employees
5 ☐ Environmental database
6 ☐ Regular measurement of environmental impact
7 ☐ Regular internal environmental reporting

5. Is there a branch-organization that provides help with environmental issues?

☐ Yes ☐ No

6. Is there a chain actor that provides help with environmental issues?

☐ Yes ☐ No

7. Please indicate participation in the following covenants.

1 ☐ None
2 ☐ Packaging Covenant
3 ☐ Energy Covenant

4 ☐ Energy Benchmarking Covenant
5 ☐ Integral Environmental Task
6 ☐ Other: ....

8. Please indicate the frequency of contacts with buyers and/or suppliers to arrange formal agreements with respect to environmental issues

1 = never / N.A., 2 = once a year, 3 = once half a year, 4 = once a trimester, 5 = once a month

	Your firm	Other firm					
1	CEO	CEO	1	2	3	4	5
2	Purchasing	Sales	1	2	3	4	5
3	Sales	Purchasing	1	2	3	4	5
4	Marketing	Marketing	1	2	3	4	5
5	R&D	R&D	1	2	3	4	5
6	Production	Production	1	2	3	4	5
7	Logistics	Logistics	1	2	3	4	5
8	Quality	Quality	1	2	3	4	5

9. Please indicate the impact of the following stakeholders on environmental management

No influence at all – 1 2 3 4 5 – Very strong influence

1	Suppliers	1	2	3	4	5
2	Buyers	1	2	3	4	5
3	Competitors	1	2	3	4	5
4	Government	1	2	3	4	5
5	Environmental organizations	1	2	3	4	5
6	Local inhabitants	1	2	3	4	5

10. To what extent do you agree with the following statements

Completely disagree –   1   2   3   4   5   – Completely agree

1	We work actively together with suppliers on environmental issues	1	2	3	4	5
2	We work actively together with buyers on environmental issues	1	2	3	4	5
3	We have a sufficient environmental budget	1	2	3	4	5

13. To what extent do you agree with the following statements

Completely disagree –   1   2   3   4   5   – Completely agree

1	Environmental information from the lower governmental agencies on demands/permits is clear	1	2	3	4	5
2	Governmental environmental policy of lower governmental bodies concerning our firm is clear	1	2	3	4	5
3	We have enough say in the environmental policy of the lower governmental agencies	1	2	3	4	5
4	There is an open dialogue with lower governmental bodies with respect to environmental policy issues	1	2	3	4	5
5	There is informal communication with lower governmental agencies on environmental policy issues	1	2	3	4	5
6	The governmental environmental policy contributes to our business goals	1	2	3	4	5

12. Please indicate the importance of a contact person at the lower government agencies to stay informed about legal environmental requirements

Not important at all –   1   2   3   4   5   – Very important

1	Visit of a local governmental civil servant	1	2	3	4	5
2	Visit to a local governmental civil servant	1	2	3	4	5
3	Fixed contact person	1	2	3	4	5

13. Please indicate the frequency of contacts with the following lower governmental agencies

1 = never / N.A., 2 = once a year, 3 = once half a year, 4 = once a trimester, 5 = once a month

1	Municipal bodies	1	2	3	4	5
2	Provincial bodies	1	2	3	4	5
3	Water authorities	1	2	3	4	5

Appendix C. Questionnaire 2005

1. What is your main position in the organization?

☐ Environmental Coordinator   ☐ CEO
☐ Quality Manager   ☐ Heath & safety manager
☐ Controller   ☐ Product Manager
☐ Board Member   ☐ Other: ....

2. To what extent do you agree with the following statements

Completely disagree –   I   2   3   4   5   – completely agree

1   Our firm wants to be first to introduce new products   I   2   3   4   5
2   Customers constantly ask for new products   I   2   3   4   5

3. Please indicate to what extent environmental pollution is a focus of attention in the organization

Not at all –   I   2   3   4   5   – Very much

1   Soil pollution   I   2   3   4   5
2   Noise   I   2   3   4   5
3   Water pollution   I   2   3   4   5
4   Smell   I   2   3   4   5
5   Air pollution   I   2   3   4   5
6   Hazardous substances   I   2   3   4   5
7   Waste production   I   2   3   4   5
8   Energy consumption   I   2   3   4   5

4. Please indicate the reasons for collecting environmental information

1 ☐ Internal environmental management   4 ☐ To check environmental emissions
2 ☐ Product (re)design   5 ☐ To evaluate environmental measures
3 ☐ Chain-oriented environmental management   6 ☐ To exchange information with buyers/ suppliers

5. Please indicate the availability of the following elements of an environmental management
   system

1 ☐ Environmental strategy
2 ☐ Environmental audit
3 ☐ Environmental action program
4 ☐ Environmental training of employees

5 ☐ Environmental database
6 ☐ Regular measurement of environmental impact
7 ☐ Regular internal environmental reporting

6. To what extent do you agree with the following statements

Completely disagree –   1   2   3   4   5   – Completely agree

1	Our organizational culture stimulates sharing environmental ideas with higher management	1	2	3	4	5
2	Environmental information can be adequately communicated to higher management	1	2	3	4	5
3	Environmental information is shared among employees	1	2	3	4	5
4	The environmentally responsible manager has a strong impact on strategic environmental issues	1	2	3	4	5
5	The environmentally responsible manager has a strong impact on operational environmental issues	1	2	3	4	5
6	Different departments are involved in environmental decision-making	1	2	3	4	5
7	Environmental problems are collectively dealt with	1	2	3	4	5
8	Board of directors is actively involved in environmental management issues	1	2	3	4	5

7. Please indicate the impact of the following stakeholders on environmental management

No influence at all –   1   2   3   4   5   – Very strong influence

1	Suppliers	1	2	3	4	5
2	Buyers	1	2	3	4	5
3	Competitors	1	2	3	4	5
4	Branch-organizations	1	2	3	4	5
5	EU	1	2	3	4	5
6	National government	1	2	3	4	5
7	Local government	1	2	3	4	5
8	Environmental organizations	1	2	3	4	5
9	Consumers	1	2	3	4	5
10	Local inhabitants	1	2	3	4	5
11	Banks and insurance firms	1	2	3	4	5

12 Employees   I   2   3   4   5

8. Please indicate participation in the following covenants

I ☐ None   4 ☐ Energy Benchmarking Covenant
2 ☐ Packaging Covenant   5 ☐ Integral Environmental Task
3 ☐ Energy Covenant   6 ☐ Other: ....

9. Please indicate the implementation of the following management systems

I ☐ Food Quality Management   3 ☐ Other: ....
2 ☐ Human Resource Management

10. To what extent do you agree with the following statements

Completely disagree –   I   2   3   4   5   – Completely agree

I	Environmental issues are integrated in quality management systems	I	2	3	4	5
2	Environmental issues are integrated in human resource management systems	I	2	3	4	5
3	We work together with suppliers on environmental issues	I	2	3	4	5
4	We work together with buyers on environmental issues	I	2	3	4	5
5	The governmental environmental policy contributes to our business goals	I	2	3	4	5

11. Please indicate to what extent attention for the environment contributed to the following items

Not at all –   I   2   3   4   5   – Very much

I	Enhanced corporate image	I	2	3	4	5	Unknown
2	Marketing opportunities	I	2	3	4	5	Unknown

Appendix D. Terms used in the questionnaires

Firm
The firm and/or plant-site/business location where you are working

Environmental manager
The manager with formal responsibilities for the coordination of environmental activities

Lower government
The lower government includes the provinces, municipalities, and the water boards

Higher government
The national government and the EU

Environmental management
All business activities and managerial and technical measures that are taken or are going to be carried out in order to prevent, reduce, and/or avoid environmental pollution and/or (other) emissions to the environment

Appendix E. Feedback report

A. Environmental management capabilities

	Mean (± S.D.)	Your firm
Level of performance	2.78 ± 1.57	4.23

0 = Lowest score; 5 = Highest score.

B. Attention for environmental issues

	Mean (± S.D.)	Your firm
Soil pollution	2.57 ± 1.31	4
Noise	3.40 ± 1.12	3
Water pollution	3.42 ± 1.27	4
Smell	3.12 ± 1.17	2
Air pollution	2.51 ± 1.21	3
Hazardous substances	2.93 ± 1.32	3
Waste production	3.28 ± 1.06	4

1 = No attention at all; 3 = Neutral; 5 = A lot of attention.

C. Organizational aspects

	Mean (± S.D.)	Your firm
Our organizational culture stimulates sharing environmental ideas with higher management	3.54 ± 0.86	3
Environmental information can be adequately communicated to higher management	3.90 ± 0.91	3
Environmental information is shared among employees	3.39 ± 1.06	4
The environmentally responsible manager has a strong impact on strategic environmental issues	3.68 ± 0.97	3
The environmentally responsible manager has a strong impact on operational environmental issues	4.01 ± 0.72	4
Different departments are involved in environmental decision-making	3.27 ± 1.15	4
Environmental problems are collectively dealt with	4.04 ± 0.91	3
Board of directors is actively involved in environmental management issues	3.77 ± 0.94	4

1 = Completely disagree; 3 = Neutral; 5 = Completely agree.

D. Integration of different quality management issues

	Mean (± S.D.)	Your firm
Environmental issues are integrated in quality management systems	3.82 ± 1.23	4
Environmental issues are integrated in human resource management systems	3.46 ± 1.22	2

1 = Completely disagree; 3 = Neutral; 5 = Completely agree.

E. Business network impact on environmental management

	Mean (± S.D.)	**Your firm**
Suppliers	2.71 ± 1.03	3
Buyers	2.89 ± 1.08	4
Competitors	2.21 ± 0.96	3
Branch-organizations	3.28 ± 0.94	4
EU	3.27 ± 0.93	3
National government	3.63 ± 0.83	3
Local government	3.83 ± 0.89	4
Environmental organizations	2.61 ± 0.99	2
Consumers	2.62 ± 1.08	2
Local inhabitants	3.33 v 1.08	4
Banks/insurers	2.60 ± 1.09	3
Employees	3.42 ± 0.86	4

1 = No influence at all; 3 = Neutral; 5 = Very strong influence.

Appendix F. Factor analyses and Cronbach α

Table E.1 and E.2 show the results of factor analyses on the variables comprising the reflective research variables (or constructs) in 2002 and 2005. The tables indicate also the Cronbach α: see Section 7.2 for a discussion on the findings.

Table E.1. Principle component factor analysis (Varimax rotation) and Cronbach α for the reflective constructs in 2002.

	Factor loadings		
Relevancy as information sources			
(Eigenvalue 2.90, Explained variance 36.3%, Cronbach α 0.75)			
Visit of governmental civil servant	**.83**	.12	.01
Visit to governmental civil servant	**.80**	-.08	.04
Fixed contact person	**.79**	.22	.04
Equality and dialogue of the governmental relationship			
(1.81, 22.6%, 0.70)			
Open-dialogue on policy goals	.11	**.81**	.27
Informal communication policy goals	.25	**.79**	.06
Enough say in environmental policy	-.08	**.66**	.28
Quality of provided governmental information			
(0.83, 10.4%, 0.69)			
Consistency of information on permits	-.01	.16	**.90**
Clear governmental environmental policy	.11	.35	**.78**

Table E.2. Principle component factor analysis (Varimax rotation) and Cronbach α for the reflective constructs in 2005.

	Factor loadings			
Environmental communication				
(Eigenvalue 3.98, Explained variance 68%, Cronbach α 0.76)				
Environmental issues can be adequately communicated to higher level management	**.79**	.22	-.11	.19
Environmental information is shared among employees	**.75**	.25	.24	.15
Organizational culture that stimulates sharing of environmental ideas	**.71**	.25	.27	.19
Influence of environmental manager				
(1.71, 19.0%, 0.78)				
Operational environmental goals and activities	.26	**.85**	.05	.15
Strategic environmental goals and activities	.45	**.76**	-.03	.17
Prospector strategy (0.76, 8.4%, 0.80)				
Customers constantly ask for new products and/or characteristics	.12	.04	**.92**	-.09
Our firm wants to be first to introduce new products	.08	.02	**.90**	.13
Involvement of different departments				
(0.62, 6.9%, 0.65)				
Different departments are involved in environmental decision making	.37	.17	-.01	**.85**
Environmental problems are collectively dealt with	.11	.60	.11	**.62**

Summary

Introduction

Attention for the natural environment (hereafter called 'the environment') has grown worldwide in recent years. A recent investigation by the Arthur D. Little Innovation High Ground Survey among 40 multinationals (Sony, Procter & Gamble, Vodafone, etc.) showed that managers expect that sustainability-driven innovation, such as environmentally product (re)design, has a growing potential to deliver value to business. However, only 5% indicated paying attention to environmental issues in their strategic planning and decision-making. These figures would probably look even worse for small- and medium-sized firms in comparison with the large multinationals. Therefore, we asked ourselves what internal and external factors, such as firm strategy and pressure from the business network (government, buyers, suppliers, local inhabitants, etc.) are influencing the attention for the environment in firms. *This book aims for a deeper understanding of the factors that have an impact on the adoption of environmental management capabilities in firms in the Dutch food and beverage industry.*

An important environmental issue in the Dutch food and beverage industry is energy consumption for heating and cooling purposes (in 2005 it accounted for approximately 8% of the total Dutch industrial energy consumption). Substantive amounts of water are also used for washing fresh products and cleaning of machinery and equipment. Many food and beverage products are suited to the consumer market, which increases the use of packaging material due to relatively small batch sizes. It is responsible for about 60% of all used packaging material (paper, glass, metal, etc.). It can be noted, though, that 60% of the used materials consist of recycled substances. Furthermore, the food and beverage industry produces organic waste. However, about 90% is re-used for different purposes, such as in animal feed (e.g. vegetables and fruit) and to generate bio-fuels.

Environmental management can be defined as all firm activities that contribute to reducing environmental impact caused by the firm's business operations. This book focuses on the commitment to the environment in terms of the adoption of environmental management capabilities. Environmental management capabilities comprise different environmental management items, such as an environmental action program, regular environmental auditing to evaluate strategic environmental targets, and an environmental database to keep record of environmental performance. In short, they reflect the capacity of the firm to take care of the environment on a structural basis. This study integrates the outside-in and the inside-out perspective. According to the outside-in perspective, the firm's competitiveness depends on industrial forces, such as rivalry among competitors and entrance of new market parties. External factors measured in this study are the different stakeholder influences from the business network, including government and societal groups as well as other chain and network actors, such as buyers, suppliers, consumers and bank and insurance firms. The inside-out perspective states that the firm's competitiveness depends on acquiring valuable resources, competences, and (dynamic) capabilities to deal with the external influences. Internal factors measured in this study are firm characteristics, including firm strategy and enabling capabilities, such as internal communication on environmental issues.

Figure 1 shows the research framework. A distinction is made between stakeholder pressures and environmental cooperation in the business network. The stakeholders can be divided into three groups: government, chain and network actors (e.g. buyers and suppliers, branch-organizations, and financial institutions), and societal groups (e.g. environmental organizations and the local community). Government can provide negative and positive incentives to increase attention for the environment. Negative incentives refer to financial sanctions and the withdrawal of environmental permits, in case a firm acts in non-compliance with environmental regulations. An example of a positive incentive is environmental support, like subsidies for investments in clean production technologies, or a less detailed environmental permit. Furthermore, government could stimulate firms to green their business through public-private environmental cooperation in terms of environmental voluntary agreements (or: covenants) between government and firms. The chain actors are buyers, suppliers, and consumers. They can exert pressure to pay attention to the environment by imposing environmental requirements, such as the use of environmentally friendly ingredients and production technologies. However, chain partners could also opt for environmental cooperation to collectively reduce environmental impact at the chain level. Important network actors are branch-organizations and financial service providers. Branch-organizations act as an intermediary between government and firms as well as between firms in the same branch (or sector). They are involved in the settlement of environmental covenants on behalf of the firms. Financial service providers include banks and insurance firms, which may impose environmental requirements conditional to the provision of loans and insurance. The societal groups consist of environmental organizations and local inhabitants, which aim to safeguard the (local) environment.

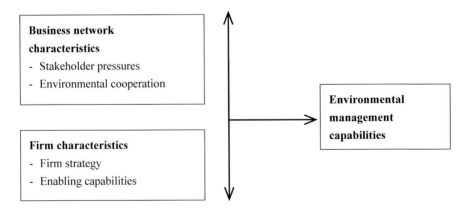

Figure 1. Research framework linking outside-in and inside-out approaches to the adoption of environmental management capabilities.

The considered firm characteristics are firm strategy and enabling capabilities (see Figure 1). The firm strategy is divided into business and environmental strategy. Business strategy is measured as the extent to which the firm is focused on innovation in business processes and products. The environmental strategy is measured as the extent to which top management shows commitment to the environment, for instance through the appointment of a board member with environmental responsibilities and, at the operational level, the appointment of an environmental manager. The enabling capabilities are divided into internal environmental feedback (e.g. the availability of sufficient internal communication channels) and the presence of already implemented food quality and safety management systems, which may support the development of environmental management capabilities. The central research question is formulated as follows. *What is the joint impact of the business network (government, buyers, suppliers, customers, etc.) and firm characteristics on the adoption of environmental management capabilities in Dutch food and beverage firms?*

Propositions

The Netherlands has a tradition of environmental covenants. Participation in environmental covenants implies that environmental measures are carried out to achieve the agreed environmental targets. Additionally, it is expected that the use of environmental covenants will increase the attention for the environment through the involvement of branch-organizations and competitors. The participation in an environmental covenant is therefore not only expected to increase attention for the environmental issues addressed in the agreement, but also to stimulate the development of environmental management capabilities in general.

P1 The participation in environmental voluntary agreements will be positively related to the adoption of environmental management capabilities.

Lower government, such as provinces and municipalities, is responsible for carrying out the governmental environmental policy. Firms have contacts with lower governmental agencies to obtain environmental permits. Contacts with lower government will therefore be important for obtaining valuable environmental information on environmental regulations and the consequences of these regulations for the firm. Although this may in particular help overcome limited resources (money and time) in small firms to take account of this, medium-sized and large firms may also benefit from contacts with lower governmental agencies.

P2 The contact intensity with lower government will be positively related to the adoption of environmental management capabilities.

Intermediary organizations, such as branch-organizations, can play an important role in increasing attention for the environment through diffusion of information on environmental measures and the translation of environmental regulations to the sector or firm level. They can also stimulate the settlement and participation in (ambitious) environmental covenants.

Furthermore, branch-organizations may increase environmental information exchange between firms.

P3 Environmental support from an intermediary, such as a branch-organization, will be positively related to the adoption of environmental management capabilities.

Chain actors, such as buyers and suppliers, will be interested in controlling environmental impacts to safeguard the proper functioning of the chain as a whole, including the regular supply of raw materials, which might be hindered in case of closure of a firm due to an environmental scandal. Large buyers in particular may therefore induce environmental standards. If a firm can not meet these standards, it will be replaced by another supplier. In summary, it is expected that increased attention for the environment among chain actors, and in particular downstream in the chain (e.g. buyers and consumers), contributes significantly to the urge to develop environmental management capabilities.

P4 Pressure from chain actors to pay attention to the environment will be positively related to the adoption of environmental management capabilities.

Firms with an innovative strategy (prospector strategy) want to be the first to introduce new products and implement new production technologies. Because it is in line with their business strategy, they are expected to be willing to invest in the environment as well, especially with respect to environmentally friendly product (re)design.

P5 A prospector strategy will be positively related to the adoption of environmental management capabilities, in particular cradle-to-cradle capabilities (e.g. environmentally friendly product (re)design).

Environmental issues are often dealt with by middle management, such as the head of the production department or plant managers (depending on the size of the firm). However, top-management commitment is essential to ensure attention for the environment at the strategic level and, consequently, the availability of sufficient resources, such as environmental budget and people, to carry out environmental activities at the operational level.

P6 Strong top-management commitment to the environment will be positively related to the adoption of environmental management capabilities.

The appointment of an environmental manager is important for increasing environmental awareness in general and to ensure that people are informed about the organizational consequences of new environmental strategies. In addition, an environmental manager can support the translation of environmental ambitions in practice as well as identify and report environmental problems.

P7 A strongly perceived impact of the environmental manager on operational and especially strategic environmental issues will be positively related to the adoption of environmental management capabilities.

Employees are important for identifying environmental problems and taking immediate care in case of environmental incidents. Their knowledge and experience is essential to successfully implement new environmental measures. Open communication at the horizontal level (such as between production, purchasing and marketing departments) and vertical level towards top-management is therefore regarding as enhancing the adoption of environmental management capabilities.

P8 Higher levels of horizontal (e.g. at the same organization level) and vertical (i.e. bottom-up and top-down) communication will be positively related to the adoption of environmental management capabilities.

Many firms in the Dutch food and beverage industry have implemented care systems to ensure food quality and safety. It is expected that the implementation of these care systems and associated working routines can be used to enable the development of environmental management capabilities.

P9 The availability of (other) care systems (e.g. food quality and safety) will be positively related to the adoption of environmental management capabilities.

Data gathering and statistical analyses

A mixed study design was employed combining semi-structured interviews involving 13 environmental managers of different firms together with two large-scale surveys carried out in 2002 and 2005 in the Dutch food and beverage industry. The interviews were carried out before and after the second survey and used to get more insight into environmental management practices. The aim of the survey questionnaires was to measure the business network influences on the attention for the environment, which formed the longitudinal backbone of the present study. However, the survey of 2002 concentrated on the relationship with government, as well. In total, 492 questionnaires could be used for the analyses, of which 386 originated from micro and small (5-50 employees) and 106 from medium-sized and large firms (50 or more employees). In 2005, special attention was paid to firm strategy and enabling capabilities. The survey included medium-sized and large firms only to get a more homogenous sample. In total, 100 questionnaires could be used. The response ratio in 2002 equaled 19% and for the medium-sized and large firms in 2002 30% and in 2005 24%. The difference between the two years can be explained by the involvement of the Dutch Ministry of Housing, Spatial Planning, and the Environment (VROM) in 2002[15].

Differences between micro, small, medium-sized and large firms are analyzed using t-tests. Furthermore, correlation and regression analyses are carried out. A cluster analysis is performed to get more insight into different firm profiles with respect to the joint impact of the business network and firm characteristics. A longitudinal analysis is carried out to get insight into changes with respect to business network impacts on the adoption of environmental management capabilities between 2002 and 2005.

Results

Environmental management capabilities

Micro and small firms have developed fewer environmental management capabilities than the medium-sized and large firms, which is understandable since they have, in general, a low environmental impact and, therefore, a low urge to pay attention to the environment. Moreover, micro and small firms are on a limited scale, which reduces the need to formally organize environmental activities. In the medium-sized and large firms, environmental management capabilities were more developed in 2005 compared to 2002. In particular, the attention for operational aspects increased, which is reflected in the fact that more firms carrying out regular checks of environmental impacts and information collection for internal environmental care. Whilst not significant, the adoption of cradle-to-cradle environmental capabilities increased, including attention for environmentally friendly product (re)design. However, overall it can be concluded that firms are focused on reducing environmental impacts at the firm level. Only a few large firms are considering chain-oriented environmental measures. For example, a respondent from a large dairy processor stated: *We have substantively improved our energy efficiency over the past few years, which makes any further improvement hard to achieve. Yet, the government still focuses on energy reduction in our chain and, therefore, we are extending our attention to suppliers to help them increase their energy efficiency as well.*

Micro and small firms

Table 1 shows the results in relation to the proposition. For the micro and small firms, no significant relationship is found between the participation in environmental covenants and the adoption of environmental capabilities *(proposition 1)*. By contrast, a significant effect is measured for the contact intensity with lower government *(proposition 2)*. The results of the correlation and regression analysis show that micro and small firms can benefit from contacts

---

[15] The survey in 2002 was carried out on behalf of the Dutch Ministry of Housing, Spatial Planning and the Environment (VROM) and paid attention to the administrative burden of environmental regulations, as well. For more information, see: Bremmers *et al.* (2003)

Table 1. Empirical assessment of the propositions for the micro, small, medium-sized and large firms.

Focus of the proposition	Micro and small	Medium-sized and large		
	2002	2002	2005	Interviews
(P1) Environmental voluntary agreements	0	0	+++	++
(P2) Contact intensity with lower government	+++	+++	a	+
(P3) Environmental support from an intermediary	0	0	a	+++
(P4) Pressure from chain actors	+	+	+++	+
(P5) Prospector strategy and cradle-to-cradle capabilities	a	a	++	++
(P6) Top-management commitment	a	a	+	+++
(P7) Perceived impact of an environmental manager	a	a	++	++
(P8) Horizontal and vertical communication	a	a	++	+++
(P9) Availability of (other) care systems	a	a	+	++

The following criteria are applied in assessing the quantitative results: +++ = significant according to the final regression model / confirmed by all respondents; ++ = significant according to a sub-model of the regression analysis or graph / confirmed by a majority of the respondents; + = significant according to the correlation analysis / confirmed by a minority of the respondents.
0 = no significant result found in the regression or correlation analysis; a = not included in the analysis.

with lower government by information exchange on environmental regulations and the consequences thereof for the firm. There is no significant relationship found for the support of an intermediary organization, such as a branch-organization *(proposition 3)*, while the perceived impact of chain actors (suppliers, buyers and competitors) is positively correlated to the development of environmental management capabilities *(proposition 4)*.

Medium-sized and large firms

Although no significant effect is measured for the participation in environmental covenants on the adoption of environmental management capabilities in 2002, a significant result is found for 2005 in the medium-sized and large firms *(proposition 1)*. Several respondents indicated obtaining valuable support from participation in covenants, although the participation itself was often perceived as semi-enforced rather than voluntary. Medium-sized and large firms appear to benefit from contacts with lower government *(proposition 2)*. The correlation analysis indicates that these contacts are, in contrast to the micro and small firms, used for environmental monitoring obligations and influencing (future) environmental regulations, though.

No significant result is found for the importance of environmental support of an intermediary organization *(proposition 3)*, whereas the significant role of branch-organizations was emphasized during the interviews. A significant result is found for the pressure of chain actors *(proposition 4)*. Especially the pressure of suppliers increased from 2002 to 2005. According to the regression analysis, suppliers have a negative impact on the adoption of environmental management capabilities. This may indicate that suppliers have, in general, less power than buyers to influence the behavior of firms, for (large) buyers can, in general, choose other suppliers. More particularly in this case, the cluster analysis shows that a large group of mainly bakery factories perceive strong pressure from their suppliers, while they are (still) lagging behind in environmental management capabilities. These bakery factories are organized in large consortia to buy ingredients collectively. In line with the expectations of an interviewee, these consortia may increasingly pay attention to the environment. Empirical evidence is found for a positive correlation between an innovative business strategy (prospector strategy) and attention for environmentally friendly product (re)design *(proposition 5)*. Commitment of top management appears to be essential to stimulate the development of environmental management capabilities *(proposition 6)*. Furthermore, the interviews and the regression analysis show that the appointment of an environmental manager *(proposition 7)* and the establishment of sufficient internal communication channels *(proposition 8)* are important too. Lastly, it turns out that working routines and procedures associated with food quality and safety assurance can be used to enhance attention for the environment *(proposition 9)*.

Discussion, conclusions and recommendations

The central research question was: *What is the joint impact of the business network (government, buyers, suppliers, customers, etc.) and firm characteristics on the adoption of environmental management capabilities in Dutch food and beverage firms?* The cluster analysis shows that an answer to this question differs for four clusters of firms. In this perspective, a cluster can be defined as a group of firms with similar business network influences and firm characteristics. One cluster of firms is characterized by a low level of environmental pollution and, consequently, minor pressure from the business network to pay attention to the environment. The most important stakeholder is lower government. Examples are medium-sized vegetables and fruit processors. Another cluster consists of firms with a medium environmental impact and weakly developed environmental management capabilities. Top management shows, however, commitment to improving environmental performance. Examples are medium-sized bakery factories, which are organized in consortia to buy ingredients collectively (as previously discussed). Another cluster consists of firms with a medium environmental impact as well, but strongly developed environmental management capabilities. Despite their attention for the environment, they perceive limited benefits due to a lack of perceived interests from buyers. Examples are medium-sized and large meat processors. The last cluster consists of firms with a high level of pollution, but also strongly developed environmental management capabilities. They perceive strong business network impacts. They have integrated environmental issues in all business activities, such as production, marketing and R&D. Examples are large dairy processors and breweries.

An important theoretical contribution of this study is the integration of the inside-out and the outside-in perspective. It appears to be a fruitful approach for increasing insight into the important factors that explain attention for the adoption of environment management capabilities. Furthermore, this study has contributed to empirical insight into the collective impact of different stakeholders on the development of environmental management capabilities. Furthermore, this study showed that qualitative and quantitative research strategies can be successfully integrated. To get more insight into longitudinal changes in stakeholder influences, it is suggested that repeated measurements of business network impacts be carried out. It would also be interesting to further investigate the relationship between a prospector strategy and the development of cradle-to-cradle capabilities to reduce environmental impacts from a product life-cycle perspective.

Concerning management and policy implications, this book showed that for micro and small firms, strong ties with government through frequent contacts with local governmental agencies are essential to stay informed about environmental regulations. The side effect of such regular contacts is that the understanding and consequently the appreciation of public environmental policy improves. Governmental agencies are therefore recommended to provide micro and small firms environmental information (e.g. through e-mails or news letters) and, more importantly, maintain the frequency of visits even if no direct monitoring or control reasons exist. The micro and small firms typically lack expertise to fundamentally change their environmental behavior. Environmental training and education programs could therefore be organized to increase the expertise to deal with environmental measures. Furthermore, cooperative pilot projects on the implementation of clean technologies can serve as a platform for knowledge exchange. It is important that intermediaries, such as branch-organizations, are involved in these initiatives, since they can take care of the administrative consequences. In the medium-sized and large firms with a defensive environmental strategy, the adoption of environmental management capabilities is most likely provoked by command-and-control regulations, while the governmental environmental policy can be differentiated for the proactive firms according to their environmental profile (see previous discussion on the results of the cluster analysis). Environmental covenants can put a strong focus on innovation addressing environmental targets at the product level (e.g. percentage of recycled components or re-usable parts) and at the chain level (e.g. chain level energy reduction targets). Large firms in particular are able to stimulate their chain partners to join them in environmental initiatives, which is essential for achieving chain-oriented environmental targets.

About the author

Derk-Jan Haverkamp (born 1979) received his MSc degree in Agricultural Engineering from Wageningen University with specializations in both Farm Technology and Business Economics (2002). Directly after his graduation he was appointed to the Management Studies Group of Wageningen University and Research Centre (WUR) as a research fellow involved in a benchmark project on the innovative capacity of four agri-food chains. He performed two evaluation projects on electronic environmental reporting on behalf of the Dutch Ministry of Housing, Spatial Planning, and the Environment (2003/2004). His PhD project started in 2004, after fulltime involvement in the organization of the 6th International Chain and Network Conference during that same year. The results of the research have been presented at (inter)national conferences and published in both academic and professional journals. Since May 2007, the author has been a consultant at ENVIRON Netherlands B.V.

Printed in the United States
by Baker & Taylor Publisher Services